2nd International Conference on Green Energy and Environment Engineering (CGEEE 2019)

ES3 Web of Conferences Volume 120 (2019)

Okinawa, Japan
2 – 5 July 2019

Editors:

Keh-Chin Chang
Jih-Hsing Chang

ISBN: 978-1-5108-9600-0

Printed from e-media with permission by:

Curran Associates, Inc.
57 Morehouse Lane
Red Hook, NY 12571

CURRAN ASSOCIATES INC. proceedings.com

Some format issues inherent in the e-media version may also appear in this print version.

For additional information, please contact EDP Sciences – Web of Conferences at the address below.

EDP Sciences – Web of Conferences
17, Avenue du Hoggar
Parc d'Activité de Courtabœuf
BP 112
F-91944 Les Ulis Cedex A
France

Phone: +33 (0) 1 69 18 75 75
Fax: +33 (0) 1 69 28 84 91

contact-edps@webofconferences.org

Additional copies of this publication are available from:

Curran Associates, Inc.
57 Morehouse Lane
Red Hook, NY 12571 USA
Phone: 845-758-0400
Fax: 845-758-2633
Email: curran@proceedings.com
Web: www.proceedings.com

TABLE OF CONTENTS

Preface

It is our great pleasure to present the volume of 2019 International Conference on Green Energy and Environment Engineering (CGEEE 2019) which was held in the Okinawa, Japan during July 2-5, 2019. The CGEEE 2019 served as a good platform for the scientific community exchanged ideas and research results, and examined the way forward regarding Green Energy and Environment Engineering. The volume tends to present to the readers the recent advances in the field of Green Energy and Environment Engineering and various related areas.

The major goal and feature of the conference is to bring academic scientists, engineers, industry researchers together to exchange and share their experiences and research results, and discuss the practical challenges encountered and the solutions adopted.

Prof. Wallace C.H. Choy from The University of Hong Kong delivered keynote speech with the title 'Low-cost Approaches for Achieving Highly Efficient and Stable Solar Cells', Prof. Miwako Hosoda from Seisa University, Tokyo, Japan had keynote speech with the title 'Condition of a Sustainable Coexistence between People and Nature', Prof. Kokyo Oh from Center for Environmental Science in Saitama, Japan delivered keynote speech 'How to Conserve Our Soil Resources from Contamination of Toxic Substances'. Besides, many researchers, engineers, academics as well as industrials from all over the world have presented their research results during the session presentations. CGEEE 2019 had six sub oral sessions with different topics: Wastewater Treatment and Bioremediation; Waste Recycling and Renewable Energy; Nanomaterials and Battery Technology; Pollution Control and Emission Reduction; Environmental Management and Satellite Applications. And two poster sessions: Energy Technology and Bioengineering; Environmental Pollution and Prevention.

We would like to thank all session chairs and the technical program committee members for their supports and dedication as always. And the gratitude also goes to all the authors who have attended the conference and contributed to this volume.

Conference Chair
Prof. Miwako Hosoda
Seisa University, Tokyo, Japan

edp sciences

E3S Web of Conferences

Statement of Peer review

In submitting conference proceedings to E3S Web of Conferences, I certify to the Publisher that I adhere to the **Policy on Publishing Integrity** of the journal in order to safeguard good scientific practice in publishing.

1. All articles have been subjected to peer review administered by the proceedings editors.
2. Reviews have been conducted by expert referees, who have been requested to provide unbiased and constructive comments aimed, whenever possible, at improving the work.
3. Proceedings editors have taken all reasonable steps to ensure the quality of the materials they publish and their decision to accept or reject a paper for publication has been based only on the merits of the work and the relevance to the journal.

Title, date and place of the conference

2019 2nd International Conference on Green Energy and Environment Engineering (CGEEE 2019)
July 2-5, 2019
Okinawa, Japan

Proceedings editor(s): Keh-Chin Chang Jih-Hsing Chang

Date and editor's signature

17, avenue du Hoggar - PA de Courtabœuf – BP 112 - 91944 Les Ulis Cedex A (France)
Tél. : 33 (0)1 69 18 75 75 - Fax : 33(0)1 69 07 45 17 – www.edpsciences.org

What is Green Manufacturing? A Quantitative Literature Review

Ira Setyaningsih[1,2,*], *Wakhid* Slamet Ciptono[2], *Nurul* Indarti[2], and *Nofie* Iman Vidya Kemal[2]

[1]UIN SUnan Kalijaga, Industrial Engineering Department, Yogyakarta, Indonesia
[2]Universitas Gadjah Mada, Management Department, Yogyakarta, Indonesia

Abstract. This paper is a decade This paper is a decade of literature review (2009-2019), about green manufacturing (GM) topic. Quantitative methods with the Publish or Perish (PoP) and VOSviewer software are used to filter the appropriate papers. The database used in PoP is Google Scholar and Scopus. During this time 403 articles were included in PoP screening, 212 articles on the Google Scholar database and 191 articles on the Scopus database. Paper not Q1 and Q2, duplicate paper in both databases, paper not in English, book review, were excluded from the analysis. The results obtained 78 papers that can be analyzed. This paper elaborate network analysis of research in green manufacturing topic and describe in detail the publications in the GM field, including journals and the year of publication. We find that there are three journals dominated the publication in this topic, Journal of Cleaner Production, International Journal of Precision Engineering and Manufacturing, and International Journal of Production Research. In this decade, GM was discussed extensively in 2018. From VosViewer software, we get four main research clusters, that is, model, green manufacturing, industry and practice. There are many topics dealing with GM can be developed in future research from maps shown by VosViewers, such as adoption, GSCM, and environmental regulation.

1 Introduction

The topic of GM emerged in the early 1990s, and was driven by human concerns about the environment [1]. Initially, GM's focus was only on process-oriented, reducing waste in the production process. This focus then develops into product-oriented, reducing resources, polluting material, and energy [1]. Green manufacturing (GM) has a variety of word similar, such as cleaner production or green practices [2]. This topic growth rapidly in this last decade. The term of green become an important and interesting topic discussion for several reason, such as the increasing of pollution and waste, the limitation and decreasing of natural resources, and global warming effect [3].

This study have several purpose, to provide an extensive literature review on green manufacturing, specifically:

- clarify the concept of GM
- overview the existing body of literature
- explore the trend in the GM research and suggest a future research topic

Some authors have their own definitions about GM. GM interpreted as paradigm, process, system or method. Despite having different explanations, the researchers agreed that GM essentially minimizes resources and reduces waste to protect the environment (Table 1.).

This paper is organized into five sections. The first section discusses GM in general and explains the purpose of the paper. In Section 2, the existing review of 'green manufacturing' definition are described. We clarify the definition of GM in this section. In Section 3 we presented the steps of the method and methodology used to conduct the literature review. The results of VOSviewer software is discusses in Section 4. Conclusions, and future research suggestion are appear in Section 5.

2. Literature Review

The literature review research on GM was examined by Rehman and Shrivastava in 2013 in sufficient detail [1]. They concluded that research on GM was classified into 12 branches, namely green design, design for environment, green technologies, life cycle assessment, green process planning, green supply chain, reverselogistic, green purchasing/marketing, green organization culture, green regulation, green productivity, green economics, and green disposal. Their paper concludes that GM includes a comprehensive set of processes in all activities in business that have an impact on the environment. Research in past, present and probably future have same direction but different focus. In the past, researchers highlighted industrial behavior that paid more attention to the final process, 'end of pipe' process, how to reduce waste or pollution. At present,

* Corresponding author: ira_darusalam@yahoo.com

researchers assess that the company's focus shifts to the 'pressure' of government or the environment to adopt GMs. Whereas in the future, it is estimated that research will focus on the life cycle and product design that are still a little highlighted [1].

Green manufacturing has been explained by various authors. Sangwan and Mittal's article on bibliometric GMs and similar frameworks found that there are eight frameworks that have been used interchangeably by researchers, for example cleaner production, sustainable production, sustainable manufacturing [2]. Dornfeld [4] defined green manufacturing as a process or system to promote sustainability and to reduce the environmental impact by minimizing resources and decreasing pollution by recycling or developing green products. Deif [5] explained that GM reflects a new manufacturing paradigm that integrate various green strategies, drivers and techniques to become more eco-efficient, including reducing material and energy, substituting input materials and recycling. Shrivastava and Shrivastava [6] noted that GM is a method for manufacturing that minimizes waste and pollution, inhibits the rate of use of natural resources and reducing input material, and reusing components, to help to build products more efficiently. Chuang and Yang [7] analyze some literature and concluded that GM concluding GM as a manufacturing method, the process of transforming materials or parts into finished goods and fulfill customer expectations that minimises waste and pollution [7]. We resume the definition by some authors in Table 1 and conclude our own definitions in Fig 1.

Table 1. GM definitions by various authors.

Author	**GM define**	**1**	**2**	**3**	**4**	**5**	**6**
Dornfeld [4]	Process; System	v	v	v	v	v	
Deif [5]	Paradigm	v	v		v		v
Shrivastava and Shrivastava [6]	Method		v	v	v		v
Chuang and Yuang [7]	Method			v			v

(1) Reduce environmental impact; (2) minimizing resources; (3) minimizing pollution; (4) recycling; (5) green products; (6) efficiently goal.

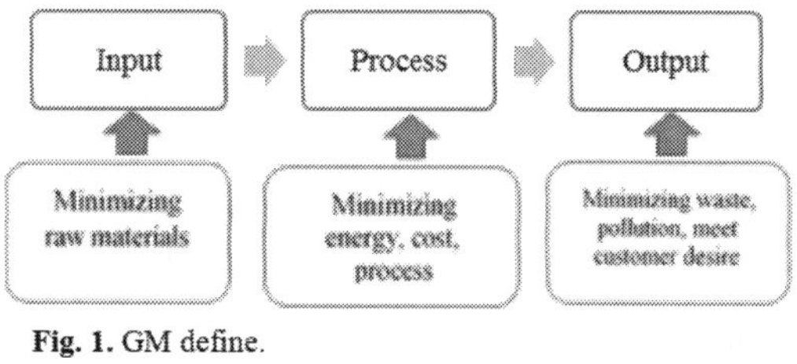

Fig. 1. GM define.

We conclude in this paper that GM is a manufacturing method with minimizing natural resources as input; process within minimal energy and cost; and produce finished goods in minimizing waste and pollution as output, but still meet customer desire (Fig. 1.).

3. Methodology

Data was collected from the Google Scholar (GS) database and Scopus database by Publish or Perish (PoP) software from Harzing [8] in Februari 2019. We used the search strings "green manufacturing" only to narrow the findings. Through this step, we get 212 papers from Scopus and 191 papers from GS. The total number in the initial search phase is 403 papers. In the next step we select and use paper from top journal (Q1 and Q2) only, based on scimago website.

Figure 2 shows many journals not analyzed. Various reasons for not analyzing this paper are presented in Table 2. Initially, we examined the paper from Scopus, this led to the Google Sholar database, 98 papers categorized as duplicated. This results in the final result, the data on google scholar seems to be only a few (four) classified into Q1 and Q2. Actually, many journals are identified in the GS database as Q1 and Q2.

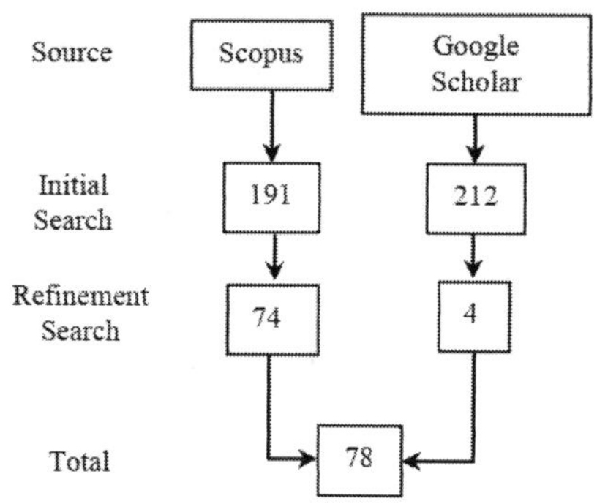

Fig. 2. Overview of paper selection process.

Figure 2 shows many journals not analyzed. Various reasons for not analyzing this paper are presented in Table 2.

Table 2. Database searching.

	Scopus	**Google Scholar**
Initial	191	212
No Article/Book Review	3	1
Not in Business/Management	37	5
Not in Scimago	12	89
Q3/Q4	65	14
Non English	0	1
Duplicate	0	98
Final Result	**74**	**4**

The highest number is paper that is not in Scimago as many as 101, followed by database duplication from Scopus and GS is 98 papers. Q3 and Q4 79 papers, and 42 papers not in business/management topic, such as green tea and green ceramics. The latest data not included in the analysis are two articles in the form of book reviews and two articles we can not obtained. Distribution of data from 2009 to 2019 from two databases, Scopus and GS, as complete as can be seen in Table 3.

Table 3. Data distribution per year.

	Scopus	GS
2009	7	12
2010	6	9
2011	9	13
2012	17	15
2013	13	21
2014	17	26
2015	25	29
2016	29	28
2017	26	18
2018	40	37
2019	2	4
Amount	**191**	**212**

Table 4. Difference of searching metric.

	Initial	**Refinement**
Papers	323	78
Citations	2479	1558
Years	10	10
Cites_Year	247,9	155,8
Cites_Paper	7,67	19,97
Cites_Author	2172,3	1466,95
Papers_Author	236,17	68,79
Authors_Paper	1,81	1,38
h_index	24	20
Cites_Author_Year	217,23	146,69
Year_first	2009	2009
Year_last	2019	2019

Data obtained from PoP software shows that there is no significant difference between Scopus and GS. From both databases, the initial and refinement metric data are obtained (Table 4). Initial metric data is data before being analyzed. In the initial data, shows the number of paper is 323, it because both databases have been merged, and the system eliminates the paper that he knows has the same data information.

Table 4 shows that the papers in Q1 and Q2 have a large impact, as seen from the increased data in citation paper, from 7.67 to 19.97.

4. Discussion

PoP software helps researcher to gain the appropriate journals. We analyze 78 journals, that spread from 2009 to February 2019. Fig 3 shows in 2018 research about GM are highest, and the lowest are in 2009.

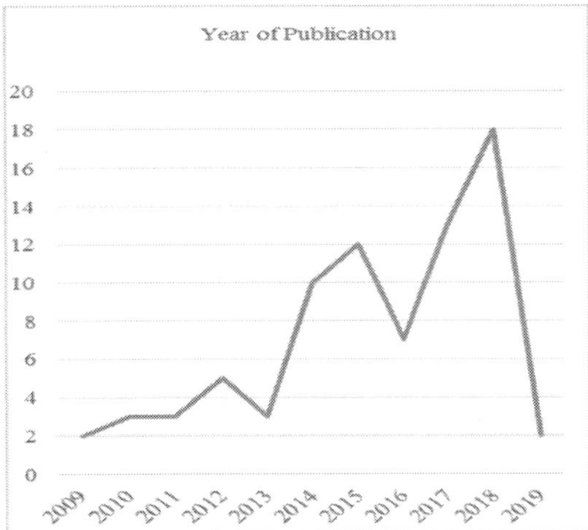

Fig. 3. Year of publication.

Figure 3 shows that in general there is an increase in research in the realm of GM topics. The decline in research occurred in 2016 and 2013. After 2016, this trend increased significantly. this means that this topic is becoming a concern in society and researchers.

The most popular Journal Publication that published this topic is Journal Cleaner Production (Table 5). This is consistent with previous research, bibliometric analysis with Google Scholar data sources only [9]. Other journals that contribute one article each and not listed in the Table 5 are International Journal of Green Economics, B.E. Journal of Economic Analysis and Policy, European Journal of Operational Research, European Journal of Wood and Wood Products, Indian Journal of Science and Technology, International Journal of Advanced Manufacturing Technology, International Journal of Business Innovation and Research, International Journal of Computer Integrated Manufacturing, International Journal of Economic Research, International Journal of Logistics Systems and Management, International Journal of Quality and Reliability Management, Journal of Environment and Development, Journal of Environmental Management, Journal of Environmental Planning and Management, Journal of Industrial Ecology, Journal of Manufacturing Systems, Journal of Remanufacturing, Journal of the Society for Information Display, World Applied Sciences, International Journal of Sustainable Development & World Ecology, International Journal of Industrial and Systems Engineering.

Table 5. Journals publication

Journal	Amount
Journal of Cleaner Production	19
International Journal of Precision Engineering and Manufacturing	8
International Journal of Production Research	8
Journal of Manufacturing Technology Management	5
Benchmarking: An International Journal	4
International Journal of Production Economics	4
International Journal of Productivity and Quality Management	2
International Journal of Procurement Management	2
International Journal of Energy Economics and	2

Policy	
International Journal of Environmental Research and Public Health	2
International Journal of Logistics Research and Applications	2
Others	20
Total	**78**

Vosviewer software that was created and explained by Van Eck and Waltman [10] was used to analyze the final data, to see the distribution of data patterns. VOSviewer shows, there are 2342 terms words that appear more than once in the title and abstract. Then the vosviewer analyzes the words, which appear above 10 times, in the 78 titles and abstracts of the papers and found 48 threshold. If we want more rigid results, we can change this amount. We can replace it with a larger number, 15 or 20 times.

Figure 4 shows density visualisation mapping from VOSViewer. The red colors presentes the largest number of items and the highest weights of elements, followed by yellow and green ones. The red colors means that these words, such as 'practice', 'model', 'process', 'performance', ' industry' are widely used by researchers. Our consideration is in the words inside green colours, such as 'energy', and ' environmental regulation'. In the middle between red and green colours, is yellow colours. The words such as 'adoption', 'survey', 'gscm', 'smes', 'cost', gives researchers an opportunity to be further investigated. Researchers can conduct future research on research topics based on these words inside green and yellow colour of density visualisation mapping.

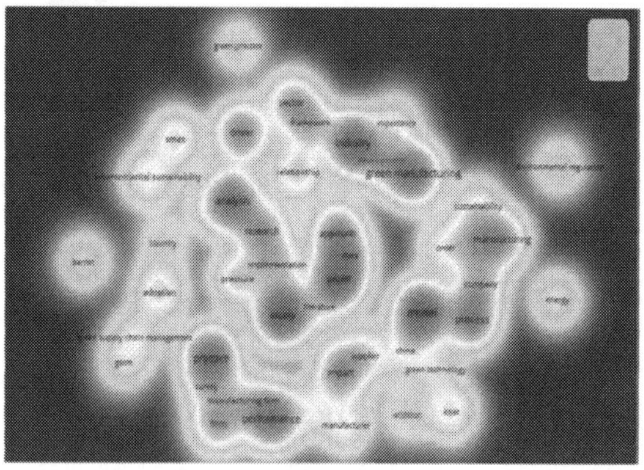

Fig. 4. Density visualisation mapping.

5. Conclusion

GM become an interesting topic discussed in line with the increasing public awareness of the importance of being friendly to the environment. People in the world are beginning to realize that the environment has limited resources and are starting to worry about its availability, such as reduced water supply and depleting supply of energy sources. The industry also began to notice about this environment issue, one of the approaches taken through adoption of GM.

GM, basically is a process of innovation from traditional manufacturing. GMs pay attention to environmental aspects in all manufacturing activities, such as inputs, processes and outputs.

In future research, there are still many opportunity to conduct research topic in GM related to adoption, SMES, GSCM, energy and environmental regulation.

Mrs. Ira Setyaningsih gratefully acknowledge to Indonesia Endowment Fund for Education (LPDP) for their support on her study and this research.

6. References

[1] Rehman, M. and Shrivastava, R. L. 2013. Green Manufacturing (GM): Past, Present and future (A State of Art Review). *World Rev. Sci. Technol. Sustain. Dev.* 10, 1/2/3, 17-55.

[2] Sangwan, K. S. and Mittal, V. K. 2015. A bibliometric analysis of green manufacturing and similar frameworks. *Manag. Environ. Qual. An Int. J.* 26, 4, 566-587.

[3] Mittal, V. K. and Sangwan, K. S. 2014. Prioritizing barriers to green manufacturing: Environmental, social and economic perspectives. *Procedia CIRP*. 17, 559–564.

[4] Dornfeld, D., Yuan, C., Diaz, N., Zhang, T., and Vijayaraghavan, A. 2013. *Introduction to Green Manufacturing*.

[5] Deif, A. M. 2011. A system model for green manufacturing. *J. Clean. Prod.* 19, 14, 1553–1559.

[6] Shrivastava, S. and Shrivastava, R. L. 2017. A systematic literature review on green manufacturing concepts in cement industries. *Int. J. Qual. Reliab. Manag.* 34, 1, 68–90.

[7] Chuang, S. and Yang, C. 2014. Key success factors when implementing a green- manufacturing system. *Prod. Plan. Control Manag. Oper.* 25, 11, 923–937.

[8] Harzing, A. W. 2011. *The publish or perish book: Your guide to effective and responsible citation analysis*, 1st ed. Melbourne, Australia: Tarma Software Research Pty Ltd.

[9] Setyaningsih, I., Indarti, N. and Jie, F. 2018. Bibliometric analysis of the term 'green manufacturing. *Int. J. Manag. Concepts Philos.* 11, 3, 315–339.

[10] Van Eck, N. J. and Waltman, L. 2010. Software survey: VOSviewer, a computer program for bibliometric mapping. *Scientometrics*. 84, 2, 523–538

A Study on the Serial Position Effect of Memory according to Illumination of LED Light

Chung Won Lee[1,*], *Jin Ho* Kim[1], and *In Keuk* Hwang[1]

[1]Kongju National University, Dept. of Industrial & Systems Engineering, Cheonan-si, Chuncheonanam-do, Republic of Korea

Abstract. The purpose of this study was to verify the forms of the effect of serial position effects of memory according to the illuminance of light. This study was conducted as an experimental method, and 21 adults without cognitive impairment participated in the study. The illuminance condition was designed with high illuminance condition of 1,000 lx and low illuminance condition of 300 lx. The memory task used word list of 20 items consisting of a series of pointless spellings. After memorizing the word list for 10 minutes, the participant performed a retention task 24 hours later. The memory retention task consisted of filling the empty part of the learned word and completing the word. The analysis was performed by dividing the word items into three conditions: primacy, middle, and recency. Primacy used the first item in the word list, Middle used the 10th item in the middle of the word list, and finally Recency used the last item in the word list for analysis. The result was F = 4.16 (p = .02), and showed that there was a statistically significant difference in memory retention of primacy, middle, and recency at 95% confidence level in dim condition.

1 Introduction

Recent studies have demonstrated that light affects memory [1], [2]. Especially, brightness of light is considered as a factor affecting memory recognition and memory recall [3]. Previous studies have reported that bright light with high illuminance is effective for working memory (short-term memory) such as attention or word processing, but light with relatively low illuminance is effective for long-term memory [3]-[5].

However, it is somewhat unreasonable to think that the relationship between illuminance and memory is a firm result. This is because unmatched research results still exist. For example, Kretschmer et al. reported that the correct response of the working memory task was excellent at a high illuminance of 3,000 lx, but reported false response was not affected by illumination [6]. On the other hand, Jung et al. argued that long-term memory is superior at relatively low illuminance of about 400 lx to high illuminance [3]. Therefore, more systematic and repeated verification of the relationship between light and memory is needed.

In order to systematically verify the effect of light on memory, there is a need to examine serial position effects of memory on light. Serial Position Effects mean that the retrieval of first and last item is better in a memory task involving a set of word lists. Excellent retrieval of an initial item is called Primacy Effect, and excellent retrieval of a last item is called Recency Effect. Generally, Primacy Effect and Recency Effect in memory retrieval are in the form of U-shaped performance curves [7][8][9]. Rundus (1971) argued that tasks in the first part of the list are more likely to be stored in long-term memories by making more rehearsals, whereas words in the last list are still short-term memories, leading to increased retrieval, and explained that this is why Serial Position Effects occur. Although this argument remains controversial, a variety of clinical studies have proven that there is a distinction between Primacy Effect and Recency Effect. As a representative example, Recency Effects of dementia patients are similar to ordinary person, but Primacy Effects are different from ordinary person [10]-[14].

Looking at Serial Position Effects, depending on the illuminance of light, we can roughly see how the overall memory retrieval pattern, including the Primacy Effect and Recency Effect. And this understanding can provide important information for exploring how differences in illuminance create differences in memory retrieval and performance. This study was conducted to investigate the characteristics of memory retrieval according to illuminance through analysis of Serial Position Effects in bright and dim conditions using LED illumination.

2 Method

2.1. Participants

This experiment was performed in adults 21 people with no cognitive impairment. The participants were 8 males and 13 females, and the mean age was 23.5 years. In addition, prior to the experiment, participants were fully

* Corresponding author: smartwony@naver.com

informed about the problems that could affect the experiment, such as alcohol or drugs, that could affect cognitive abilities.

2.2 Lighting Environment

The LED lighting used was the GM10743 model lighting manufactured by Ningbo Golemore Industrial Co., Ltd. This product has a diameter of 9 cm and 5 small LED bulbs are inserted. For this experiment, several lightings were used to construct illuminance conditions. In this study, the illuminance condition was designed with high illuminance condition of 1,000 lx and low illuminance condition of 300 lx. The color temperature was 5,500k and the spectral power distribution was the same as the figure.

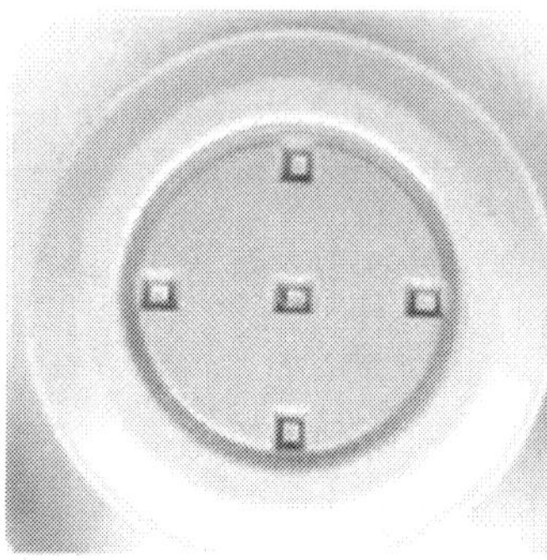

Fig. 1. LED lighting(GM10743).

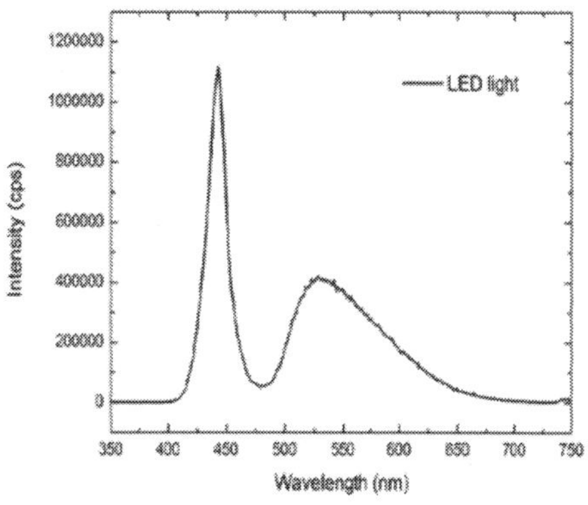

Fig. 2. Spectral power distribution.

2.3 Memory Task

The memory task used word list of 20 items consisting of a series of pointless spellings. After memorizing the word list for 10 minutes, the participant performed a retention task 24 hours later. The memory retention task consisted of filling the empty part of the learned word and completing the word.

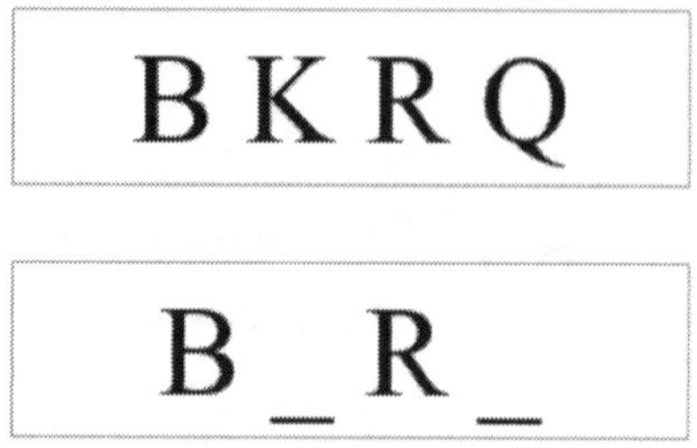

Fig. 3. Example of memory task.

2.4 Experimental Procedure

The experiment was designed with a dim condition of 300 lx and a bright condition of 1,000 lx. For the experiment, the participants first underwent dark adaptation for 2 minutes. After 2 minutes of light adaptation, the participants memorized 20 words items over 10 minutes. After 24 hours from the completion of the memory task, the memory retention was measured through the retention task.

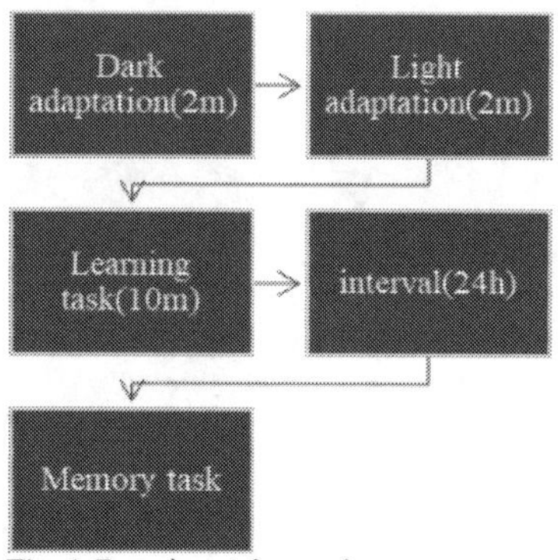

Fig. 4. Experimental procedure.

3 Result

The analysis was performed by dividing the word items into three conditions: primacy, middle, and recency. Primacy used the first item in the word list, Middle used the 10th item in the middle of the word list, and finally Recency used the last item in the word list for analysis. Memory retention was measured as 1 if the memorized word was correctly completed, and 0 otherwise.

The results of descriptive statistics were as follows.

Table 1. Descriptive statistics.

	dim condition(300lx)		bright condition(1000lx)		N
	Mean	SD	Mean	SD	
primacy	.76	.44	.71	.46	21
middle	.38	.50	.52	.33	

recency	.71	.46	.43	.51	

Then, the variance analysis was performed to verify whether serial position effect appears in the dim condition. The result was F = 4.16 (p = .02), and showed that there was a statistically significant difference in memory retention of primacy, middle, and recency at 95% confidence level.

Table 2. Result of mean difference (dim condition).

(I)	(J)	Mean difference (I-J)	.sig
primacy	middle	.38*	.028
primacy	recency	.05	.941
recency	middle	.33	.061

**p<.05, **p<.01*

As a result of the post hoc test, the retention mean difference between primacy and middle items is. 38, indicating a statistically significant difference at the 95% confidence level. However, no significant difference was found between primacy and recency, middle and recency. In dim conditions, U-shape curve shape, which is a general aspect of serial position effect, appeared.

However, no statistically significant difference was found between primacy and middle, primacy and recency, middle and recency in bright conditions. In addition, the U-shape curve due to serial position effect was not shown in general, and the retention decreased as the session progressed.

Table 3. Result of t-test (bright condition).

	dim condition	bright condition	.sig
primacy	.76	.71	.733
middle	.38	.52	.279
recency	.71	.49	.064

**p<.05, **p<.01*

Additionally, the difference verification between primacy, middle, and recency according to illuminance conditions was performed. As a result, there was no significant difference between the dim condition and the bright condition. However, although not statistically significant, the pattern of serial position effects showed a different pattern. Especially in bright condition, recency effect was very low compared with dim condition.

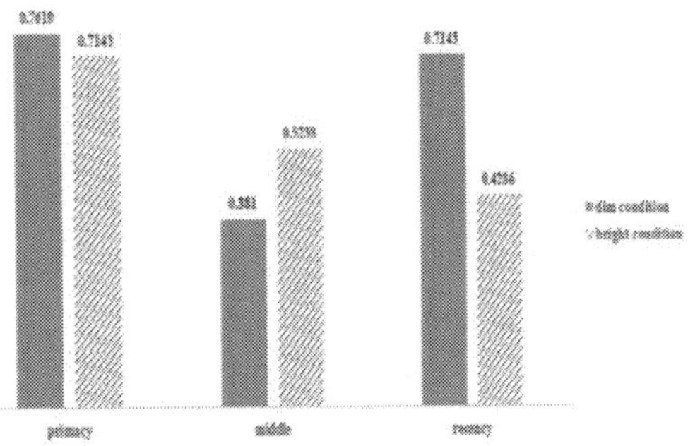

Fig. 5. Aspect of serial position effects by light condition.

4 Conclusion and Discussion

The purpose of this study was to verify the forms of the effect of serial position effects of memory according to the illuminance of light. The study verified that primacy effect appears in both bright and dim conditions where the brightness of light is relatively dark. And the results of this study showed that the difference between primacy and middle items was statistically significant in relatively dim conditions. These results show that the primacy effect is strong in the dim condition, and the retrieval is low in the middle item. The recency effect was also good in the dim condition, but it was low in the relatively bright condition.

One of the possibilities that the recency effect does not appear well under bright light conditions may be due to the depletion of attentiveness over time due to the increased use of attention in bright light conditions compared with dim conditions. Previous studies have consistently shown that attention is high in bright light conditions rather than in dark light conditions. Therefore, in order to verify this aspect of recency effect, it is necessary to consider both the attention aspect and fatigue in light conditions. This is a research topic to be covered in future research.

Despite the contribution of research, this study has limitations in that primacy, middle, and recency were measured only as one item and the difficulty of item was not taken into consideration. These limitations may be helpful for a more systematic understanding of the effects of light on memory if supplemented by further studies.

5 Acknowledgement

This work was supported by the National Research Foundation of Korea(NRF) grant funded by the Korea government(MSIT). (No. 2018R1D1A1B07050211))

References

1. Badia, P., Myers, B., Boecker, M., and Culpepper, J. 1991. Bright light effects on body temperature, alertness, EEG and behavior. Physiol.Behav. 50, 582-588.

2. Cajochen, C. 2007. Alerting effects of light. Sleep Med. Rev. 11(6), 453–464.
3. Jung, H. C., Kim, J. H., and Lee, C. W. 2017. The effect of the illuminance of light emitting diode (LED) lamps on long-term memory. Displays. 49, 1-5.
4. Chellappa, S. L., Gordijn, M.C.M., and Cajochen, C. 2011. Can light make us bright? Effects of light on cognition and sleep. Prog. Brain Res, 190, 119–133.
5. Chang, A. M., Santhi, N., St Hilaire, M., Gronfier, C., Bradstreet, D. S., Duffy, J. F., Lockley, S. W., Kronauer, R. E., and Czeisler, C. A. 2012. Human responses to bright light of different durations. J. Physiol. (Lond.). 590(13), 3103–3112.
6. Kretschmer, V.,Schmidt, K., and Griefahn, B. 2012. Bright light effects on working memory, sustained attention and concentration of elderly night shift workers, Light. Res. Technol. 44(3), 316–333.
7. G.A Carlesimo, M Sabbadini, L Fadda, and C Caltagirone. 1997. Word-list forgetting in young and elderly subjects: evidence for age-related decline in transferring information from transitory to permanent memory condition. Cortex, 33, 155-166.
8. F.I.M Craik. 1970. The fate of primacy items in free recall. Journal of Verbal Learning and Verbal Behavior. 9, 143-148.
9. Rundus, D. 1971. Analysis of rehearsal processes in free recall. Journal of Experimental Psychology. 89, 63-77.
10. Baddeley, A.D., and Warrington, E.K. 1970. Amnesia and the distinction between long- and short-term memory. Journal of Verbal Learning and Verbal Behavior. 9 ,176-189.
11. Bigler, E.D., Rosa, L., Schultz, F., Hall, S., and Harris, J. 1989. Rey-Auditory Verbal Learning and Rey-Osterrieth Complex Figure design performance in Alzheimer's disease and closed head injury. Journal of Clinical Psychology, 45, 277-280.
12. Burkart, M., Heun, R., and Benkert, O. 1998. Serial position effects in dementia of the Alzheimer type Dementia and Geriatric. Cognitive Disorders. 9,130-136.
13. Carlesimo, G.A., Sabbadini, M., Fadda, L., and Caltagirone, C. 1995. Different components in word-list forgetting of pure amnesics, degenerative demented and healthy subjects. Cortex, 31,735-745.
14. Gibson, A.J. 1981.A further analysis of memory loss in dementia and depression in the elderly. British Journal of Clinical Psychology, 20,179-185.A. Mecke, I. Lee, J.R. Baker jr., M.M. Banaszak Holl, B.G. Orr, Eur. Phys. J. E **14**, 7 (2004)

E3S Web of Conferences **120**, 01003 (2019)
CGEEE 2019
https://doi.org/10.1051/e3sconf/201912001003

Prediction of Density and Speed of Sound of Binary Ionic Liquid and Ketone Mixtures Using Artificial Neural Network

Redden Rose Rivera[1,*], and *Allan* Soriano[2]

[1]Mapua University, School of Chemical Engineering and Chemistry, 1002 Manila, Philippines
[2]De Lasalle University, Gokongwei College of Engineering, Chemical Engineering Department, 2401 Manila, Philippines

Abstract. The applications of ionic liquids solve a lot of major problems regarding green energy production and environment. Ionic liquids are solvents used as alternative to unfriendly traditional and hazardous solvents which reduces the negative impact to environment to a great extent. This study produced models to predict two of the basic physical properties of binary ionic liquid and ketone mixtures: density and speed of sound. The artificial neural network algorithm was used to predict these properties by varying the temperature, mole fraction, atom count in cation, methyl group count in cation, atom count in anion, hydrogen atom count in anion of ionic liquid and atom count in ketone. Total experimental data points of 2517 for density and 947 for speed of sound were used to train the algorithm and to test the network obtained. The optimum neural network structure determined for density and speed of sound of binary ionic liquid and ketone mixtures were 7-9-9-1 and 7-7-4-1 respectively; overall average percentage error of 2.45% and 2.17% respectively; and mean absolute error of 28.21 kg/m3 and 33.91 m/s respectively. The said algorithm was found applicable for the prediction of density and speed of sound of binary ionic liquid and ketone mixtures.

1 Introduction

Ionic liquids (ILs) are classified as highly useful compounds having very good properties such as being non-volatile and non-flammable, low toxicity and good solubility for various organic and inorganic materials. Compared to regular volatile organic solvents, ILs have received considerable attention because they are more environmentally friendly. Applications of ILs solve a lot of major problems regarding green energy production and environment which includes solar energy, biomass and CO_2 adsorption [1]. ILs have also been introduced as carbon dioxide capturing solvents as alternative to conventional amine-based solvents [2]. Publications about the preparation and application of ILs have grown and researches have gained considerable interest, therefore physical properties of ILs such as density, viscosity, conductivity, speed of sound, etc. are needed for these applications.

ILs are considered as "designer solvents" and an alternative to hazardous organic solvents [3]. The use of these promising solvents as alternative to unfriendly traditional solvents reduces the negative impact on environment to a great extent. On the other hand, the toxic effects of ILs also have attracted great attention. In order to control their potential hazards and to design environmentally friendly ILs, knowing its physical properties is highly important. The numerous applications of ILs encourage the need to have thermodynamic models to compute the thermodynamic properties of ILs like the COSMO-RS which have been widely-used and considered an efficient tool in predicting properties of ILs [4]. Previous studies are only limited to certain number of systems of ionic liquids like binary mixtures of ILs with water and alcohol. Also, they only vary only one, two or three parameters like pressure, temperature and mole fraction while the effect of atom counts is disregarded. Other researches considered a certain temperature range but limited binary ILs only.

Model building is needed for the properties of ILs to integrate wide range of composition and temperature as varying parameters. The existing data from experiments are reviewed to achieve accurate predictions of ILs. This research generates models that can predict physical properties of binary IL and ketone's physical properties by incorporating the parameters that are produced from an algorithm. These properties are specifically density and speed of sound which are acquired from experimental data from different references. Optimization is done using the algorithm namely Artificial Neural Network (ANN).

The output of this research is useful in reducing the cost in terms of utilizing ILs. Conducting experiments is highly costly because of the expensive materials and equipment. In this research, the physical properties are predicted instead of utilizing experiments which minimizes expenses and time. The huge number of experimental data can be represented by a model so that industries that use ILs will no longer handle experiments but will only refer to the models that have the ability to

* Corresponding author: reddenrose@gmail.com

https://doi.org/10.1051/e3sconf/201912001003

predict properties. This makes the experimental data simpler and manageable.

This study covers the prediction of binary IL and ketone mixtures only. The ketones considered are acetone, acetophenone, gamma-butyrolactone, N-methylpyrrolidone and 4-methyl-1,3-dioxolan-2-one. The ANN algorithm is used to predict the properties of ILs namely density and speed of sound. Density is an important property of ionic liquids which is essential in different applications like designing apparatuses, applying and solving material balance and design of separation processes and a crucial property in industrial process design involving ionic liquids. Speed of sound is also a significant property because it is needed to determine other physical properties and it is one of the key thermodynamic property of ILs due to its close relation with compressibility. The IUPAC Ionic Liquid Database, ILThermo is the database used in obtaining the experimental data of ILs. The parameters that are varied are temperature, mole fraction, atom count in cation, methyl group count in cation, atom count in anion, hydrogen atom count in anion of IL and atom count in ketone only.

2 Methodology

The step by step process flow chart for the prediction of density and speed of sound using the Artificial Neural Network algorithm is shown in Figure 1. The first step was gathering all the available experimental data for binary IL and ketone mixtures from different references. Inconsistent data were then trimmed.

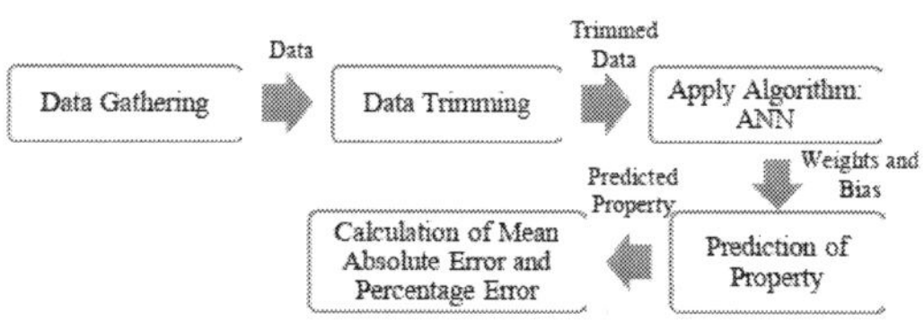

Fig. 1. Conceptual framework.

The acquired trimmed data were considered as the input data in evaluating the weights and bias or simply called as the parameters by applying the Artificial Neural Network algorithm. Using the obtained values of the weights and bias, equations for computing the properties used by the algorithm were utilized to recalculate the values of the properties. These new set of predicted data acquired after applying the model was then tested by getting the mean absolute error between the predicted and experimental values in order to identify which combination of the hidden layers produced the most accurate results. The values of experimental and predicted properties were organized and then graphed according to the value of the percentage error.

2.1 Data gathering

Data gathering involved the searching for property data of binary IL and ketone mixtures from a database which is the IUPAC Ionic Liquids Database, ILThermo. The codes of cations and anions used in this study is presented in Table 1. A total of 2565 for density and 947 experimental data points for speed of sound were gathered. The temperature range, pressure, mole fraction range, number of data points and references for the respective ILs were also presented in the discussion. Also, information regarding atom count in cation, atom count in anion, methyl group count in cation, hydrogen atom count in anion of IL, and atom count in ketone were gathered, compiled and organized completely in a Microsoft Excel® file as part of data warehousing.

Table 1. Cation and anion codes of ionic liquids.

Cation Codes	Anion Codes
(C1) 1-ethyl-3-methylimidazolium	**(A1)** Ethyl sulfate
(C2) tetrabutylphosphonium	**(A2)** Bromide
(C3) 1-hexyl-3-methylimidazolium	**(A3)** Tetrafluoroborate
(C4) phosphonium, tetraethyl-,	**(A4)** Trifluoromethane Sulfonate
(C5) 1-butyl-3-methylimidazolium	**(A5)** 1,1,2,2,2-pentafluoro-N-[(pentafluoroethyl)sulfonyl] Ethanesulfonamide
(C6) Ethylammonium	**(A6)** bis[(trifluoromethyl) sulfonyl]imide
(C7) 1-methyl-3-propylimidazolium	**(A7)** Iodide
(C8) Tetraphenylphosphorane	**(A8)** Hexafluorophosphate
(C9) Propylammonium	**(A9)** Nitrate
(C10) Triethylammonium	**(A10)** Formate
(C11) Trimethylammonium	**(A11)** Acetate
(C12) Diethylammonium	**(A12)** Dihydrogen phosphate
(C13) Tetraethylammonium	**(A13)** Hydrogen sulfate
(C14) Tetrapropylammonium	
(C15) Tetrahexylammonium	
(C16) 1-butyl-1-methylpyrrolidinium	

2.2 Data trimming

Data trimming was the collating of all the data gathered by observing them very carefully, and then removing the points of disagreement. Accurate and reliable data was considered. Data trimming was done by graphing all the data gathered from different references. Graphs of temperature versus property and mole fraction versus property were plotted that formed smooth curves. Points that were observed to be far or out from the smooth curve were omitted or trimmed. The trimmed data were compiled into one spread sheet in preparation for the utilization of the algorithm. For density, the number of data points left after trimming was 2517 and 947 for speed of sound.

References were classified into three different groups namely systems with more than two references, systems with two available references and systems with only one reference. For systems with more than two references, the data that were consistent with one another were considered. For systems with two available references, the most accurate data were selected in which the accuracy were based from the reported uncertainty. Lastly, systems with only one available reference were automatically considered. Additionally, systems that contain only two data points that belongs to this group were considered [5]. Data with large uncertainties were inconsistent data and expected to be out from the generated smooth curve.

2.3 ANN algorithm

The ANN model used in the algorithm is presented in Figure 2. This model includes input to a layer succeeded by hidden layers and the last layer called the output layer [6]. Each node in all layers were bridged through weights ($w_{n-n'}$, $w_{I'-n}$, and $w_{n'-Y}$) and also the bias (w_{b-n}, $w_{b-n'}$, and w_{b-Y}). The source node is at first subscript while the destination node is at second subscript.

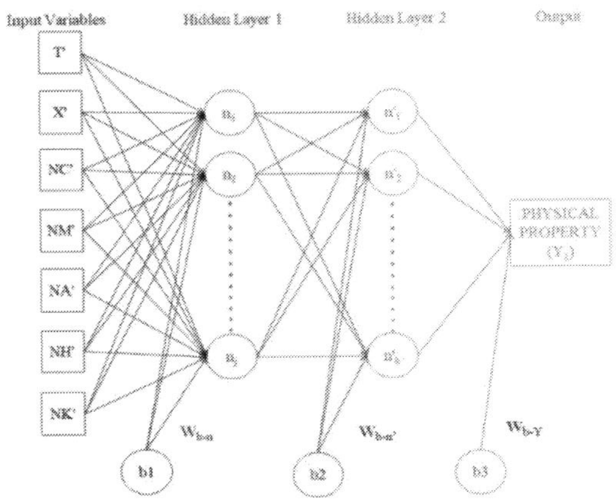

Fig. 2. ANN model with two hidden layers.

The setting of input layer allowed a total of seven input variables: temperature, T, mole fraction of the IL, X, atom count in cation, NC, CH_3-R group count in cation, NM, atom count in anion, NA, hydrogen atom count in anion, NH of IL and atom count in ketone, NK. The study was limited to a maximum of 2 hidden layers and 10 nodes in each hidden layer. For the output layer, there was only one expected output per neural network structure which was the physical property. Density and speed of sound for each binary IL and ketone mixture were treated separately.

A flowchart for determining the neural network structure is presented in Figure 3. The neural network structure with the most accurate results for every physical property was determined by selecting the structure with lowest mean absolute error, MAE given in Eq. 1, where L is the total number of data points, Y_l is the predicted value and E_l is the experimental or actual value.

$$\text{MAE}=\frac{1}{L}\sum_{l=1}^{L}\left|Y_l-E_l\right| \quad (1)$$

The mathematical relationship among the inputs, hidden layers and outputs is described by weights, bias weights, and transfer functions [7]:

$$n_j = \text{f}\left(\sum_{j=1}^{J} I' \times w_{I-n} + 1 \times w_{b-n}\right) \quad (2)$$

$$n'_k = \text{f}\left(\sum_{k=1}^{K'} n_j \times w_{n-n'} + 1 \times w_{b-n'}\right) \quad (3)$$

$$Y_l' = \text{f}\left(\sum_{l=1}^{L'} n'_k \times w_{n'-Y} + 1 \times w_{b-Y}\right) \quad (4)$$

The transfer function is given by the following:

$$\text{f}_1 = \text{logsig} = \frac{1}{1+e^{-x}} \quad (5)$$

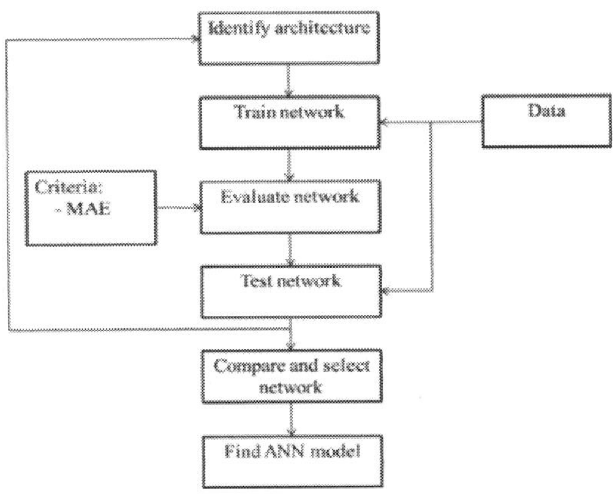

Fig. 3. Flowchart for determining neural network structure.

The I' in Eq. 6 is a function of the input variables: T, X, NC, NM, NA, NH and NK and its corresponding slope, m_i and intercept, b_i and is expressed as:

$$I' = m_i I + b_i \quad (6)$$

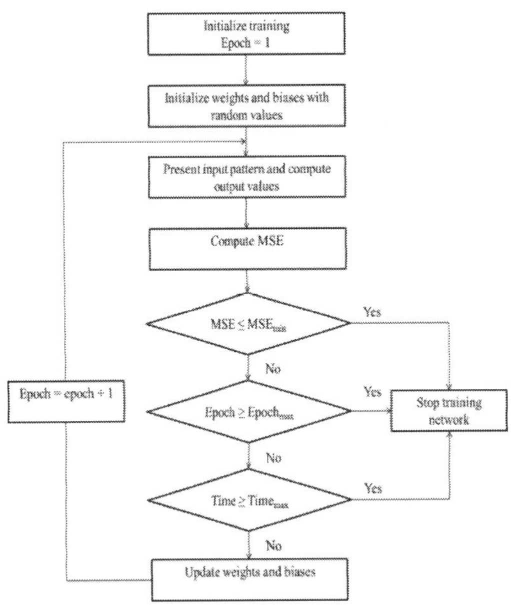

Fig. 4. Training process flowchart.

Initially, the training of ANN started by randomly assigning the initial values for weights and bias. A flowchart for training the neural network is shown in Figure 4. The weights and bias were adjusted until the termination criteria was achieved during the training period. In this case, a value of 0.5 for all initial weights and bias were assigned. The maximum number of epochs, training time and target mean square error, MSE (Eq. 7) were the three termination criteria.

$$MSE = \frac{1}{L}\sum_{l=1}^{L}(Y_l - E_l)^2 \quad (7)$$

The experimental data were divided into two different subsets which were the training and testing. Ninety percent (90%) of the experimental data were utilized in neural network training and remaining ten percent (10%) of data were allocated in testing the generated model. Parameters were obtained, and the optimum neural network structure were selected after the training process.

2.4 Prediction of properties

An MS Excel® spread sheet which simplified the computation for the chosen structure was prepared for the recalculation of the properties. Eq. 9 was used to recalculate the data points for density and speed of sound. Y_l is the predicted value and Y_l' is the output node and is a function of weights, bias weights and the corresponding slope and intercept for the given set of data points (Eq. 4).

$$Y_l' = m_l Y_l + b_l \quad (8)$$

$$Y_l = \frac{Y_l' - b_l}{m_l} \quad (9)$$

2.6 Computation of MAE and percentage error

The last step was obtaining the mean absolute error for density and speed of sound using Eq. 1. For the goal of analyzing the results, percentage errors given in Eq. 10 were also computed and the experimental and predicted values were grouped and graphed according to the percentage errors. Percentage errors of 0 – 1% were marked green, 1 – 5% were marked yellow while the 5% and above were marked red.

$$\text{Percentage Error} = \frac{|E_l - Y_l|}{E_l} \times 100\% \quad (10)$$

3 Results and discussion

3.1 Density

A total of 2565 data points was collected for the density of binary IL and ketone mixtures from different reliable literatures. From that, 2517 trimmed data were utilized in training the network and obtaining the optimum parameters of the neural network. Adjustment of parameters was performed for different trial-and-error stage or epoch until the optimum neural network structure was achieved. The optimum neural network structure in the prediction of density of binary mixture of IL and ketones is called the 7-9-9-1 neural network structure. The neural network with nine (9) hidden layer 1 nodes and nine (9) hidden layer 2 nodes was the combination that reported the lowest mean absolute error of 28.21 kg/m^3 and overall average percentage error of 2.45%.

Table 2. Density prediction of studied binary mixture of IL and acetone.

IL Code	Temp. Range (K)	Pressure (kPa)	Mole Fraction Range	Exp. Density Range (kg/m^3)	Ave. Percent Error	Data Points	Ref.
C1A1	278.15-308.15	101	0.0000-1.0000	773.58-1251	1.31	140	[8]
C1A3	293.15-308.15	101	0.1214-0.9203	918.15-1275.5	1.74	72	[9]
C2A2	288.15-313.15	101	0.0000-0.0166	766.8-819.6	1.23	102	[10]
C5A3	298.15	101	0.0000-0.0018	784.6-787.1	3.63	8	[11]
C1A4	278.15-318.15	101	0.0500-0.9501	844.2-1390.5	2.60	65	[12]
C4A7	298	101.3	0.0000-0.0008	784.5-786.1	2.44	18	[13]
C5A8	298.15	101.3	0.0000-1.0000	784.9-1367.9	3.54	15	[14]
C3A8	298.15	101	0.0000-1.0000	784.4-1293.7	2.20	14	[15]
C3A6	288.15-298.15	101.3	0.0000-1.0000	784.3-1381.1	2.91	39	[16]
C1A5	298.15-313.15	101	0.0000-1.0000	766.9-1593.2	3.18	42	[17]
Total					1.96	515	

Table 3. Density prediction of studied binary mixture of IL and gamma-butyrolactone.

IL Code	Temp. Range (K)	Pressure (kPa)	Mole Fraction Range	Exp. Density Range (kg/m^3)	Ave. Percent Error	Data Points	Ref.
C8A2	288.15-313.15	101.3	0.0023-0.0104	1111.9-1144	1.20	60	[18]
C7A6	293.15-323.15	100	0.0000-1.0000	1098.9-1479.3	1.32	84	[19]
C5A6	293.15-323.15	100	0.0000-1.0000	1098.9-1439.3	0.66	112	[20]
C16A6	293.15-323.15	100	0.0000-1.0000	1098.9-1399	0.50	126	[21]
C6A9	293.15-318.15	101	0.0000-1.0000	1096.7-1210.2	1.47	108	[22]
Total					0.98	490	

Table 4. Density prediction of studied binary mixture of IL and N-methylpyrrolidone.

IL Code	Temp. Range (K)	Pressure (kPa)	Mole Fraction Range	Exp. Density Range (kg/m^3)	Ave. Percent Error	No. of Data	Ref.
C9A10	293.15-333.15	100	0.0952-1.0000	974.8-1031.1	3.60	99	[23]
C9A11	293.15-333.15	100	0.1090-1.0000	964.6-1028.3	4.88	99	[23]
C10A13	298.15-313.15	101	0.0000-1.0000	1014.7-1142.9	1.28	60	[23]

IL Code	Temp. Range (K)	Pressure (kPa)	Mole Fraction Range	Exp. Density Range (kg/m^3)	Ave. Percent Error	No. of Data	Ref.
C12A11	298.15-313.15	101	0.0000-1.0000	1010.1-1030.2	3.75	60	[24]
C10A11	298.15-313.15	101	0.0000-1.0000	1000.3-1026.9	4.99	60	[24]
C12A13	298.15-313.15	101	0.0000-1.0000	1014.7-1284	5.29	60	[24]
C10A12	298.15-313.15	101	0.0000-1.0000	1014.7-1125.7	1.20	64	[25]
C11A13	298.15-313.15	101	0.0000-1.0000	1014.7-1467.6	6.17	64	[25]
C11A11	298.15-313.15	101	0.0000-1.0000	1014.7-1053.9	5.02	60	[25]
C11A12	298.15-313.15	101	0.0000-1.0000	1014.7-1353.6	6.99	60	[25]
C5A3	298.15-333.15	101	0.0000-1.0000	997-1200.6	2.43	105	[26-28]
C5A6	298.15-323.15	100	0.0000-1.0000	1004.8-1433.7	1.38	66	[29]
C5A8	298.15-313.15	101	0.0953-0.8938	1083.5-1351.1	6.82	36	[25]
C1A3	293.15-308.15	101	0.1113-0.9304	1070.6-1276.8	5.73	72	[30]
C6A9	298.15	101	0.0000-1.0000	1028.3-1210.8	2.14	20	[31]
C1A6	293.15-323.15	101	0.0978-0.9002	1115.9-1502.2	1.24	63	[32]
Total					3.89	1048	

Table 5. Density prediction of studied binary mixture of IL and 4-methyl-1,3-dioxolan-2-one.

IL Code	Temp. Range (K)	Pressure (kPa)	Mole Fraction Range	Exp. Density Range (kg/m^3)	Ave. Percent Error	Data Points	Ref.
C5A3	298.15	101	0.0000-0.8497	1199.6-1206.5	1.30	15	[33-34]
C1A1	278.15-308.15	101	0.0000-1.0000	1188.9-1251	2.25	133	[8]
C13A3	283.15-313.15	101	0.0000-0.0826	1177.8-1215	0.75	62	[35]
C5A6	293.15-323.15	101	0.0146-0.8980	1183.3-1432.8	0.73	98	[36]
C14A2	298.15	101	0.0013-0.0102	1197.3-1199.1	0.19	7	[37]
C16A6	293.15-318.15	101	0.0000-1.0000	1200.2-1387	1.18	144	[37]
C15A2	298.15	101	0.0097-0.0550	1193.1-1198.1	1.30	5	[37]
Total					1.32	464	

As shown in Tables 2 to 5, the average percentage error for the prediction of density of binary IL and ketone mixtures ranged from 1.23% to 3.63%, 0.50% to 1.47%, 1.20% to 6.99% and 0.19% to 2.25% for binary ILs containing acetone, acetophenone, gamma-butyrolactone and 4-methyl-1,3-dioxolan-2-one respectively. The results acquired from prediction show that N-methylpyrrolidone + trimethylammonium dihydrogen phosphate (C11A12) with 60 data points resulted the highest percentage error. The possible sources of errors for the systems with a bit high percentage error could be from the inconsistency of experimental data that might occurred due to impurities of the substances and from inaccuracy of instruments and measurements methods.

Tables 6, 7 and 8 show the values of the weights and bias that resulted from applying the ANN algorithm that are used in determining A, Bn' and Cn of Eq. 11. The values are the result of modeling the density of binary IL and ketone mixtures which obtained the 7-9-9-1 as the optimum neural network structure and were utilized in predicting density. The generated model for this prediction as shown in Eq. 11 is a function of parameters A, Bn' and Cn where mole fraction, X, temperature, T, atom count in cation, NC, methyl group count in cation, NM, atom count in anion, NA, hydrogen atom count in the anion, NH and atom count in the ketone, NK values are inputted in Cn.

$$\text{Predicted Density} = \frac{1}{0.0012}\left(\frac{1}{1+e^{-A}}\right) + 766.799 \quad (11)$$

Where:

$$A = \sum_{n'=1}^{9}\left[W_{b-Y} + W_{n'-Y}\left(\frac{1}{1+e^{-B_{n'}}}\right)\right]$$

$$B_{n'} = \sum_{n=1}^{9}\left[W_{b-n'} + W_{n-n'}\left(\frac{1}{1+e^{-C_n}}\right)\right]$$

$$C_n = W_{b-n} + 0.0182W_{T-n}(T - 278.15)$$
$$+W_{x-n}X + 0.0152W_{NC-n}(NC - 11)$$
$$+0.0161W_{NM-n} + 0.05W_{NA-n}(NA - 1)$$
$$+0.0161W_{NH-n}NH$$
$$+0.1667W_{NK-n}(NK - 10)$$

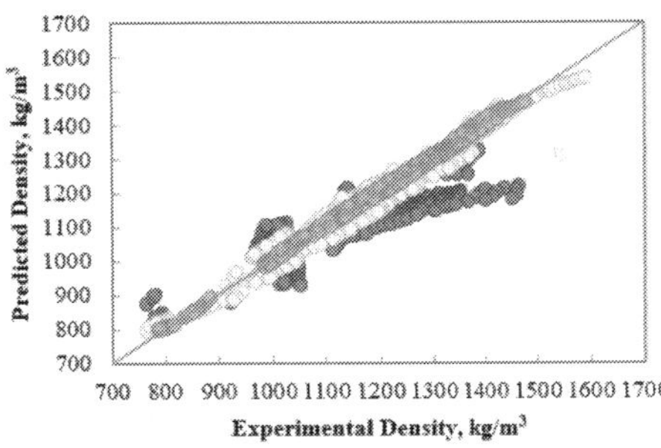

Fig. 5. Density of binary IL and ketone mixtures.
Predicted values by thes model versus experimental values: %error < 1 (Green); 1 ≤ %error ≤ 5 (Yellow); %error > 5 (Red).

Figure 5 shows that the optimum neural network structure was able to predict the experimental values successfully for the density of binary and ketone mixtures. The predicted values were observed to have good agreement with experimental values. Some of the error of prediction lied in the experimental uncertainties of data points. From all the recalculated experimental density of this system, 34.92% were in between the percentage error range of 0-1% which were marked green, 50.74% were in the range of 1-5% which were marked yellow and only 14.34 % of the studied compounds have the percentage error higher than 5% which were marked red. With such minimal number of data points with percentage error greater than 5%, the model obtained using the 7-9-9-1 structure was already acceptable for the prediction of density of binary IL and ketone mixtures.

Table 6. Hidden layer 1 weights and bias.

n	W_{b-n}	W_{T-n}	W_{X-n}	W_{NC-n}	W_{NM-n}	W_{NA-n}	W_{NH-n}	W_{NK-n}
1	-0.1758	-0.5749	-0.1919	-0.1422	1.5232	-0.1299	0.5558	-0.4187
2	-0.7970	-0.3672	-0.2888	0.1891	-0.4901	-0.1126	-0.4911	-0.1316
3	-0.6827	-0.0926	4.9775	0.2008	-0.5951	-0.7771	-0.4150	5.6131
4	4.1191	0.4785	-2.6178	6.2242	-2.8601	-3.1228	11.7474	-0.7126
5	0.7235	-0.5036	7.3794	-1.1647	-1.5854	0.6500	-1.1372	7.0075
6	-0.8060	-0.6615	0.3296	-0.3239	3.0801	0.0101	0.1983	-0.3027
7	-2.7877	0.4201	1.9049	-2.4259	-5.1720	-0.5317	3.7196	0.6978
8	-0.7899	-0.3609	1.5800	-2.3205	-2.1130	-2.2754	10.3176	0.5380
9	-4.3872	0.7050	-3.8286	0.4812	-2.8008	0.6431	-0.9270	3.9505

Table 7. Hidden layer 2 weights and bias.

n'	$W_{b-n'}$	$W_{1-n'}$	$W_{2-n'}$	$W_{3-n'}$	$W_{4-n'}$	$W_{5-n'}$	$W_{6-n'}$	$W_{7-n'}$	$W_{8-n'}$	$W_{9-n'}$
1'	-0.6568	-0.2366	-0.3948	0.5680	-1.0631	1.0366	-0.2419	-0.0288	-1.2717	-0.2266
2'	-0.3633	0.0953	0.2219	-0.4847	0.5201	-0.9987	0.0900	-0.2120	0.4470	0.4504
3'	-0.2613	-0.5905	0.0711	0.6762	-1.9827	0.7623	0.0337	-0.3381	-0.5887	-0.9828
4'	0.1062	-0.0106	-0.1498	-1.2979	1.5211	-2.5178	-0.3335	0.7707	0.7357	0.8750
5'	-0.1991	-0.2184	-0.1846	0.6696	-1.6270	0.5756	0.3878	-0.6201	-0.7753	-0.6732
6'	-0.6594	0.3852	-0.3986	0.1549	-1.3091	0.8048	0.2691	-0.6612	-0.0411	-0.2894
7'	-0.7074	0.1990	-0.3667	-0.0475	-0.2679	-0.2821	-0.3917	0.2404	0.0454	-0.2681
8'	-0.0848	-1.0879	0.0418	-1.6752	2.0314	-2.0672	-1.0718	1.9989	1.8753	1.7967
9'	0.0338	-0.3481	-0.2178	-0.7598	0.4263	-0.4517	-0.4303	-0.2501	0.4338	0.6150

Table 8. Output layer weights and bias.

W_{b-Y}	$W_{1'-Y}$	$W_{2'-Y}$	$W_{3'-Y}$	$W_{4'-Y}$	$W_{5'-Y}$	$W_{6'-Y}$	$W_{7'-Y}$	$W_{8'-Y}$	$W_{9'-Y}$
0.0382	1.2361	-0.9824	1.7933	-3.1517	1.5534	1.0979	-0.0532	-3.7370	-0.8666

3.2 Speed of sound

There were 947 data points collected for the speed of sound of binary IL and ketone mixtures which were obtained from various reliable literatures. Since all the systems have only one reference, all the collected data were used in training the network and obtaining the optimum parameters of the neural network. Just like in the prediction of density, adjustment of parameters was performed for different trial-and-error stage or epoch until the optimum neural network structure was achieved. The optimum neural network structure in predicting the speed of sound of binary IL and ketone mixtures is the 7-7-4-1 neural network structure. The structure with seven (7) hidden layer 1 nodes and four (4) hidden layer 2 nodes achieved the smallest mean absolute error of 33.91 m/s and 2.17% overall average percentage error.

Table 9. Speed of sound prediction of studied binary mixture of IL and acetone.

IL CODE	Temp. Range (K)	Pressure (kPa)	Mole Fraction Range	Exp. Speed of Sound Range (m/s)	Ave. Percent Error	Data Points	Ref.
C1A3	293.15-308.15	101	0.1214-0.9203	1172.8-1614.6	1.96	72	[7]
C1A4	278.15-318.15	101	0.0500-0.9489	1109.3-1475.9	1.70	65	[10]
C3A6	288.15-298.15	101.3	0.0000-1.0000	1161.6-1250.5	0.79	39	[14]
C2A2	298.15	101	0.0006-0.0166	1161.12-1178.7	1.08	13	[8]
Total					1.57	189	

Table 10. Speed of sound prediction of studied binary mixture of IL and N-methylpyrrolidone.

	Temp. Range (K)	Pressure (kPa)	Mole Fraction Range	Exp. Speed of Sound (m/s)	Ave. Percent Error	Data Points	Ref.
C12A11	298.15-313.15	101	0.0000-1.0000	1454-1617	2.64	60	[22]
C9A10	293.15-333.15	100	0.0952-1.0000	1420.3-1566	1.64	99	[21]
C1A3	293.15-308.15	101	0.1113-0.9304	1520.3-1621.3	0.63	72	[28]
C11A13	298.15-313.15	101	0.0000-1.0000	1520-1591	3.48	64	[23]
C10A12	298.15-313.15	101	0.0000-1.0000	1520-1836	2.34	64	[23]
C9A11	293.15-333.15	100	0.1090-1.0000	1391.1-1558.1	0.59	99	[21]
C10A11	298.15-313.15	101	0.0000-1.0000	1520-1840	2.24	60	[22]
C12A13	298.15-313.15	101	0.0000-1.0000	1356-1572	3.34	60	[22]
C10A13	298.15-313.15	101	0.0000-1.0000	1520-1894	3.66	60	[22]
C11A11	298.15-313.15	101	0.0000-1.0000	1490-1586	2.73	60	[23]
C11A12	298.15-313.15	101	0.0000-1.0000	1520-1708	4.01	60	[23]
Total					2.32	758	

The percentage error for the prediction of the speed of sound of binary IL and ketone mixtures ranged from 0.79% to 1.96% and 0.59% to 4.01% for ILs containing acetone and N-methylpyrrolidone respectively, as presented in Tables 9 and 10. The results acquired from prediction show that N-methylpyrrolidone +

Table 11. Hidden layer 1 weights and bias.

n	W_{b-n}	W_{T-n}	W_{X-n}	W_{NC-n}	W_{NM-n}	W_{NA-n}	W_{NH-n}	W_{NK-n}
1	0.3946	0.6883	2.0038	-4.8584	0.1144	-4.0348	-0.5635	4.7758
2	0.8904	-2.0383	-0.6031	2.0653	1.3870	-0.3993	-0.0254	1.6676
3	-2.4562	1.8953	2.1869	-4.0276	-5.1550	2.4945	-2.0634	-1.1748
4	0.2491	0.2349	1.1831	-3.7918	0.3209	-2.3343	-0.5430	3.9458
5	1.5601	-0.5615	-2.1579	-12.3581	-0.5566	-1.9587	2.8208	4.1721
6	-0.6752	-1.1269	0.2481	1.2642	0.7233	-0.2572	-0.2764	-0.2656
7	-0.8933	2.8635	-3.7984	2.1791	-0.2367	-1.6760	0.4127	-1.3612

Table 12. Hidden layer 2 weights and bias.

n'	$W_{b-n'}$	$W_{1-n'}$	$W_{2-n'}$	$W_{3-n'}$	$W_{4-n'}$	$W_{5-n'}$	$W_{6-n'}$	$W_{7-n'}$
1'	1.3502	-2.2289	-0.4700	2.5172	-1.3798	3.9218	0.0744	0.9831
2'	0.0448	-1.9471	-0.3800	0.6523	-1.1998	0.9135	0.2236	1.6648
3'	-0.2964	-0.9532	-0.6120	-0.0231	-0.6493	0.7089	0.2225	0.2593
4'	0.3143	0.8527	0.2445	-0.5045	0.3972	-0.2384	0.3080	-0.4534

Table 13. Output layer weights and bias.

W_{b-Y}	$W_{1'-Y}$	$W_{2'-Y}$	$W_{3'-Y}$	$W_{4'-Y}$
2.8065	-4.2321	-2.7572	-0.9407	1.7635

trimethylammonium dihydrogen phosphate (C11A12) with 60 data points resulted the highest percentage error.

Tables 11, 12 and 13 show the values of weights and bias that resulted from applying the ANN algorithm that are used in determining A, Bn' and Cn of Eq. 12 to predict the speed of sound. The generated model for the prediction of this system given in Eq. 12 is a function of parameters A, Bn' and Cn where T, X, NC, NM, NA, NH and NK values were inputted in Cn.

$$\text{Predicted Speed of Sound} = \frac{1}{0.0013}\left(\frac{1}{1+e^{-A}}\right) + 1109.3 \quad (12)$$

Where:

$$A = \sum_{n'=1}^{4}\left[W_{b-Y} + W_{n'-Y}\left(\frac{1}{1+e^{-B_{n'}}}\right)\right]$$

$$B_{n'} = \sum_{n=1}^{7}\left[W_{b-n'} + W_{n-n'}\left(\frac{1}{1+e^{-C_n}}\right)\right]$$

$$C_n = W_{b-n} + 0.0182W_{T-n}(T - 278.15)$$
$$+W_{x-n}X + 0.0256W_{NC-n}(NC - 14)$$
$$+0.0164W_{NM-n}(NM - 1) + 0.0714W_{NA-n}(NA - 1)$$
$$+0.0161W_{NH-n}NH + 0.1667W_{NK-n}(NK - 10)$$

Figure 6. shows that the optimum neural network structure was able to predict the experimental values successfully. The predicted values of speed of sound of binary IL and ketone mixtures were observed to have good agreement with experimental values. Most of the error of prediction lied in the experimental uncertainties of data points. From all the recalculated experimental speed of sound of this system, 35.06% were in between the percentage error range of 0-1% which were marked green, 53.01% were in the range of 1-5% which were marked yellow and only 11.93% were greater than 5% which were marked red. With such minimal percentage error greater than 5%, the model obtained using the 7-7-4-1 structure was already acceptable for the prediction of speed of sound of binary IL and ketone mixtures.

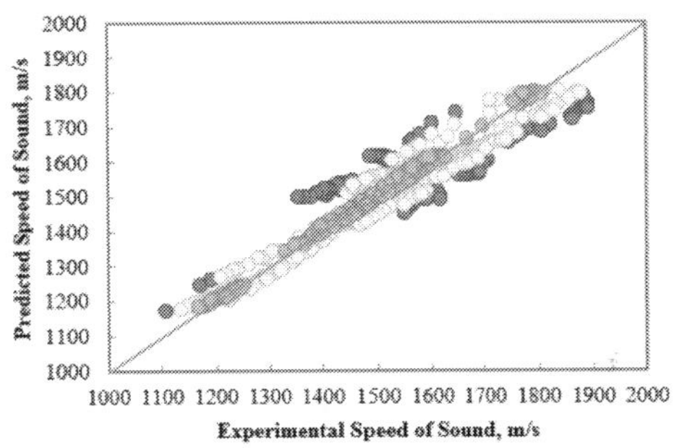

Fig. 6. Speed of sound of binary IL and ketone mixtures. Predicted values by the model versus experimental values: %error < 1 (Green); 1 ≤ %error ≤ 5 (Yellow); %error > 5 (Red).

4 Conclusion

New models using an algorithm, the artificial neural network, were designed to predict the density and speed of sound of binary ionic liquid and ketone mixtures which have numerous applications because of its environmentally benign nature. The prediction was based on the knowledge of temperature, mole fraction, atom count in cation, methyl group count in cation, atom count in anion, hydrogen atom count in anion of IL and atom count in ketone for the compounds in the mixture. Density and speed of sound from IUPAC Ionic Liquids Database, ILThermo, were considered as the experimental database. If no experimental data is available, the proposed models are potential tools to estimate the density and speed of sound of binary ionic liquid and ketone mixtures.

In this study, the network has a maximum of seven inputs and one output. The optimum neural network structure acquired for the density of binary ionic liquid and ketone mixtures was the 7-9-9-1 model with an overall average percentage error of 2.45% and mean absolute error of 28.21 kg/m^3. For the speed of sound of binary ionic liquid and ketone mixtures, the model with 7-7-4-1 neural network structure was found to be the optimum model which produced 2.17% overall average percentage error with 33.91 m/s mean absolute error. The results for the density and the speed of sound of binary ionic liquid and ketone mixtures showed that the experimental and the ANN predicted data were in good agreement. Therefore, the identified models are applicable for the prediction of density and speed of sound of binary IL and ketone mixtures.

References

1. T. Z. and M. F. Feng Guo, S. Zhang, J. Wang, Botao Teng, Curr. Org. Chem. **19**, 455 (2015)
2. F. K. Chong, N. Chemmangattuvalappil, F. Eljak, M. Atilhan, D. Foo, Designing ionic liquid solvents for carbon capture using property-based visual approach, **18** (2016)
3. P. L. Amado Alviz, A. J. Alvarez, J. Clean. Prod. **168**, 1614–1624 (2017)
4. M. Diedenhofen, A. Klamt, Fluid Phase Equilib. **294** (1–2), 31–38 (2010)
5. A. N. Soriano, A. M. Agapito, L. J. L. I. Lagumbay, A. R. Caparanga, M. H. Li, J. Taiwan Inst. Chem. Eng. **41** (3), 307–314 (2010)
6. E. Y. Li, Artificial neural networks and their business applications, Inf. Manag. **27** (5), 303–313 (1994)
7. V. D. Nguyen, R. R. Tan, Y. Brondial, T. Fuchino, Prediction of vapor-liquid equilibrium data for ternary systems using artificial neural networks, Fluid Phase Equilib. **254** (1–2), 188–197 (2007)
8. A. P. F. J. Lehmann, M. H. Rausch, A. Leipertz, J. Chem. Eng. Data **55** (9), 4068–4074 (2010)
9. S. Bhagour, S. Solanki, N. Hooda, D. Sharma, V. K. Sharma, J. Chem. Thermodyn. **60**, 76–86 (2013)
10. J. Krakowiak, W. Grzybkowski, J. Chem. Eng. Data **55** (7), 2624–2629 (2010)
11. V. Kiselev, H. A. Kashaeva, I. I. Shakirova, L. N. Potapova, A. Konovalov, Solvent Effect on the Enthalpy of Solution and Partial Molar Volume of the Ionic Liquid 1-Butyl-3-methylimidazolium Tetrafluoroborate, **41** (2012)
12. E. Vercher, F. J. Llopis, M. V. González-Alfaro, A. Martínez-Andreu, J. Chem. Eng. Data, **55** (3), 1377–1388 (2010)
13. L. Pinheiro, A. Calado, J. Carlos R. Reis, C. A. N. Viana, *Shape Effects in the Partial Molar Volume of Tetraethylphosphonium and Ammonium Iodides in Six Nonaqueous Solvents*, **32** (2003)
14. J. Wang, A. Zhu, Y. Zhao, K. Zhuo, *Excess Molar Volumes and Excess Logarithm Viscosities for Binary Mixtures of the Ionic Liquid 1-Butyl-3-methylimidazolium Hexaflurophosphate with Some Organic Compounds*, **34** (2005)
15. A. Pereiro, E. Tojo, A. Rodriguez, J. Canosa, J. Tojo, *Properties of Ionic Liquid HMIMPF6 With Carbonates, Ketones and Alkyl Acetates*, **38** (2006)
16. E. González, A. Domínguez, E. A. MacEdo, *Excess properties of binary mixtures containing 1-hexyl-3-methylimidazolium bis(trifluoromethylsulfonyl)imide ionic liquid and polar organic compounds*, **47** (2012)
17. N. Deenadayalu, P. Bhujrajh, *Density, speed of sound, and derived thermodynamic properties of ionic liquids [EMIM](+)[BETI](-) or ([EMIM](+)[CH(3)(OCH(2)CH(2))(2)OSO(3)](-) plus methanol or plus acetone) at T = (298.15 or 303.15 or 313.15) K*, **53** (2008)
18. J. Krakowiak, J. Wawer, A. Farmas, *Apparent molar volumes and compressibilities of electrolytes and ions in gamma-butyrolactone*, **54** (2012)
19. S. Papović, M. Vranes, S. Gadžurić, *A comprehensive study of {γ-butyrolactone+1-methyl-3-propylimidazolium bis(trifluoromethylsulfonyl)imide} binary mixtures*, **91** (2015)
20. M. Vranes, S. Papović, A. Tot, N. Zec, S. Gadžurić, *Density, excess properties, electrical conductivity and viscosity of 1-butyl-3-methylimidazolium bis(trifluoromethylsulfonyl)imide + gamma-butyrolactone binary mixtures*, **76** (2014)
21. M. Vranes, S. Papović, A. Tot, N. Zec, S. Belic, S. Gadžurić, *Ideal and non ideal behavior of 1-butyl-1-methylpyrrolydinium bis(trifluoromethylsulfonyl)imide + gamma-butyrolactone binary mixtures*, **81** (2014)
22. R. Zarrougui, M. Dhahbi, D. Lemordant, *Volumetric Properties of Ethylammonium Nitrate + γ-Butyrolactone Binary Systems: Solvation Phenomena from Density and Raman Spectroscopy*, **39** (2010)
23. P. Chhotaray, R. Gardas, S. Jella, *Structural and compositional effect on the acoustic and volumetric properties of ammonium based ionic liquids with water and N-methyl-2-pyrrolidone*, **219** (2016)
24. T. Kavitha, P. Attri, P. Venkatesu, R. S. Rama Devi, and T. Hofman, *Temperature dependence measurements and molecular interactions for ammonium ionic liquid with N-methyl-2-pyrrolidone*, **54** (2012)
25. T. Kavitha, P. Attri, P. Venkatesu, R. S. Rama Devi, T. Hofman, *Influence of temperature on thermophysical properties of ammonium ionic liquids with N-methyl-2-pyrrolidone*, **545** (2012)
26. F. Qi and H. Wang, *Application of Prigogine–Flory–Patterson theory to excess molar volume of mixtures of 1-butyl-3-methylimidazolium ionic liquids with N-methyl-2-pyrrolidinone*, **41** (2009)
27. S. Tian, Y. Hou, W. Wu, S. Ren, K. Pang, J. Chem. Eng. Data **57** (3), 756–763 (2012)
28. J.Y. Wu, Y.P. Chen, and C.S. Su, *Density and Viscosity of Ionic Liquid Binary Mixtures of 1-n-Butyl-3-methylimidazolium Tetrafluoroborate with Acetonitrile, N,N-Dimethylacetamide, Methanol, and N-Methyl-2-pyrrolidone*, **44** (2014)
29. T. Srinivasa Krishna, N. Kolla, M. Sankar, A. Nain, B. Munibhadrayya, *Thermodynamic, excess and optical studies on the intermolecular interactions of binary liquid mixtures of imidazolium based ILs*, **98** (2016)
30. D. Sharma, S. Bhagour, V. K. Sharma, *Thermodynamic and Topological Studies of 1-Ethyl-3-methylimidazolium Tetrafluoroborate + Pyrrolidin-2-one and 1-Methyl-pyrrolidin-2-one Mixtures*, **57** (2012)
31. M. Usula, E. Matteoli, F. Leonelli, F. Mocci, F. Marincola, L. Gontrani, S. Porcedda, *Thermo-physical properties of ammonium-based ionic liquid + N-methyl-2-pyrrolidone mixtures at 298.15 K*, **383** (2014)
32. H. Yao, S. Zhang, J. Wang, Q. Zhou, H. Dong, and X. P. Zhang, *Densities and Viscosities of the Binary Mixtures of 1-Ethyl-3-methylimidazolium Bis(trifluoromethylsulfonyl)imide with N-Methyl-2-pyrrolidone or Ethanol at T = (293.15 to 323.15) K*, **57** (2012)
33. I. Bou Malham, M. Turmine, J. Chem. Thermodyn. **40** (4), 718–723 (2008)
34. J. H. and R. B. Stoppa, A., J. Chem. Eng. Data **54**, 472–479 (2009)
35. E. Tyunina, V. Afanas'ev, M. Chekunova, *Viscosity and Density of Solutions of Tetraethylammonium Tetrafluoroborate in Propylene Carbonate at Different Temperatures*, **41** (2012)
36. N. Zec, A. Tot, S. Papović, S. Belic, M. Vranes, S. Gadžurić, *Density, electrical conductivity, viscosity and excess properties of 1-butyl-3-methylimidazolium bis(trifluoromethylsulfonyl)imide + propylene carbonate binary mixtures*, **68** (2014)

37. Y. Zhao, J. Wang, X. Xuan, R. Lin, *Volumetric studies of ion solvation in propylene carbonate + N,N-dimethylformamide electrolyte solutions*, **81** (2003)

A Study on Protective Reuse of Traditional Architecture in Lhasa

Qing Qin[1,*], *Yang* Chen[1], and *Haiyue* Zhang[1]

[1]School of Human Settlements and Civil Engineering, Xi'an Jiaotong University, Xi'an 710049, China

Abstract. Traditional Tibetan architecture in Lhasa has a unique position in the world. Under the special social background of economic development demand and tourism, it has gradually become a pillar industry, and how to rationally protect and reuse this traditional architecture has become an urgent problem that must be solved. Based on the traditional architectural patterns, regional characteristics and modern needs of Lhasa, this paper discusses the strategies for the protection and reuse of architectural heritage, the specific methods of reuse and their advantages and disadvantages, in order to provide a reference for the protection of traditional buildings and the inheritance of culture in Tibet.

1 Introduction

Traditional Tibetan architecture, as one of the world's most unique architectural types, has distinctive characteristics and occupies a very special position. Under the combined effect of the particularity of the natural environment caused by the snow plateau and the humanistic environment formed by the Tibetan people's religious beliefs, traditional Tibetan architecture has gradually formed its unique architectural system and design methods. Lhasa has been the political, economic and cultural center of Tibet since ancient times. The population of Lhasa accounts for about 27% of the total population of Tibet, and has the most representative and well-preserved Tibetan architecture. Therefore, this study on the protective reuse of traditional buildings in Lhasa can provide a theoretical basis and strategies for the protection of traditional buildings in Tibet.

2 Overview of traditional architecture in Lhasa

2.1. The types and characteristics of traditional architecture in Lhasa

Lhasa is located on the Tibetan plateau at an altitude of more than 3500 meters, forming a plateau temperate semi-arid monsoon climate, which is cold and dry in winter and cool and rainless in summer. Under the influence of this climatic condition, the unique flat-roofed building form with a wooden framework and mud and stone wall structure has developed in Lhasa. Traditional Tibetan architecture generally uses wooden beams and columns as load-bearing systems in the interior and mud or stone walls as load-bearing walls in the exterior. Most of the buildings do not have more than three storeys, and their plane shape is mostly rectangular. In Lhasa, "L" and "U"-shapes are the most common plane forms of traditional residential buildings, and the layout is relatively compact. In order to minimize heat loss, most spaces are closely connected with the kitchen, where the stove is located. The kitchen also assumes the function of a living room. The habits of people living around fire makes the two functional spaces of traditional Tibetan architecture overlap. Frequently used rooms, such as the living room and the main bedroom, are usually arranged in the south, while a large number of secondary bedrooms and bathrooms are placed in the north to maximize the exposure to sunlight of the frequently used rooms. On the facade, the most notable feature of Tibetan architecture is that the wall is gradually retracted upwards. The building's structure uses thicker walls at the bottom, which taper off a bit at the top. This kind of wall does not only increase the stability of the walls, but is also very effective at retaining heat. Apart from the walls, the windows above are larger than the ones below to prevent heat loss. In terms of the vertical organization of building functions, the less important functional space is usually placed at the bottom, such as rooms for livestock or storage.

In addition to the influence of natural environmental factors on the formation of traditional Tibetan architecture in Lhasa, social and human environmental factors are also very important. Influenced by the widespread influence of Tibetan Buddhism, the Buddhist room, as an important and frequently used space in Tibetan daily life, is usually arranged on the top floor of the building. In terms of architectural decoration, the exterior walls are mostly white, and white pulp is regularly sprayed on the walls. In addition to maintaining the whiteness of the wall, it also has become a part of religious customs. The parapet walls, eaves and portals of doors and windows are usually black, which plays the role of warding off evil spirits and absorbing heat. The interior decoration usually uses red, white, yellow, green, and blue patterns with religious meanings. (Fig. 1)

* Corresponding author: chenyang@xjtu.edu.cn, 710516568@qq.com

Fig. 1. Traditional buildings of Lhasa.

2.2 The current situation of traditional architecture in Lhasa

The existing traditional Tibetan architecture in Lhasa can be roughly divided into religious buildings, large manor buildings that are scattered throughout the Lhasa urban area, and other buildings closely arranged in Barkhor (Bakuo) Street in the old part of Lhasa's urban area, as well as temples and mansions. Apart from the temples, which still have their original religious functions, most of the other buildings have now been converted into scenic spots, or commercial or residential buildings. As a typical residential building in Tibet, the large manor buildings in Lhasa have become a hot spot for tourism development and have been mostly transformed into homestay or hotel buildings. After the government recovered the traditional Tibetan buildings in Barkhor Street, some of them were given to civilians or the descendants of the aristocrats who originally lived there, while some of them were rented out as housing. After some of these buildings were privately purchased, others became private residences or were sub-leased to merchants from all around the country. The lower floors of these buildings are used for business purposes such as operating various types of individual shops or small workshops, while the upper floors are used for accommodation. Some other buildings were transformed into hotels, teahouses, homestays, hotels, medium-sized shops and so on. The function of traditional Tibetan architecture in Lhasa has changed throughout history. In order to meet the needs of modern society, most of this architecture has become residential or commercial buildings.

3 The significance and problems of protective reuse of traditional architecture in Lhasa

3.1 The significance and value of protective reuse

The overall social level of Tibet has entered a new stage due to the continuous progress and development of the society. As the political, economic and cultural center of Tibet, Lhasa not only is responsible for promoting economic development, but should also prioritize the protecting and inheriting of Tibet's history and culture. As an important carrier of Tibetan history, culture and religious beliefs, traditional Tibetan architecture plays an important role in reflecting the development of the city. In the process of protecting traditional Tibetan architecture, reuse as a means of restoring the original function of buildings is an ideal one among many methods for protecting ancient buildings. The reuse of traditional architecture can not only discover its role as a witness to the social and economic development of each historical period, but can also extend its protection from that of the architecture itself to the surrounding environment and cities, and even that of history, culture, and people's current life. It enables people to truly gain a sense of the typical social environment and rich local historical and cultural folk traditions, and to explore a deeper comprehensive protection of the human environment and cultural characteristics.

At the same time, the rapidly developing tourism industry has become a major pillar of Lhasa's economy. In this context, we should consider the problem of how buildings can better serve the economy. With the development of cities and the tourism industry, there is an increasing demand and market for traditional buildings to be used for commercial activities. Integrating and coordinating the protective reuse and commercial reuse of traditional buildings is one of the most effective means of meeting the needs of multiple parties.

3.2 Problems in the process of reuse

The biggest problem in the process of the reuse of existing traditional buildings in Lhasa is the lack of professional guidance and protection. The reuse of most buildings is only to simply change their function, but this method neglects its historical and social value. At the same time, in order to meet the needs of modern life, it is extremely common for the current residents to renovate the interior space of buildings. Due to the lack of professional knowledge, the phenomenon of self-construction is more serious, which not only changes the spatial structure and use nature of traditional buildings, but also has great potential safety hazards. Because of users' limited awareness of protection, the ancient buildings are constantly damaged during the process of renovation, and the original traditional architectural style is destroyed.

From the user's point of view, the spatial organization form of traditional buildings may not be able to meet the needs of modern life. Many buildings have problems such as poor ventilation and lighting. In addition, since the traditional Tibetan buildings in Lhasa mostly use a wooden frame as an internal load-bearing system, the government generally does not allow private heating equipment to be added during the renovation process, resulting in the building having a low level of indoor comfort during the winter, which cannot meet basic usage and living needs.

4 Strategies and methods of protective reuse of traditional architecture in Lhasa

 https://doi.org/10.1051/e3sconf/201912001004

4.1 Reuse strategy combining regional characteristics and modern demand

The reuse of traditional buildings in Lhasa should be based on the preservation status of buildings, the surrounding environment and resources, and the functional orientation of the future development of Lhasa. On this basis, it is necessary to meet the living needs of modern residents and users and maximize the utilization of social resources. Before reusing traditional buildings, an overall evaluation and analysis of the traditional buildings and even the entire block should be carried out, and the protection and renewal should be combined during the process of reuse. For culturally significant architecture in Lhasa, such as temples and palaces, it is necessary to repair and renovate the original buildings to the greatest extent, and restore the original features of the building. The newly built buildings around them which are not in harmony with the traditional buildings should be demolished. For traditional buildings that are well preserved but are not culturally significant architecture, their original architectural functions can be retained or restored. The interior space of the buildings can be updated and rebuilt according to their functions and users, and the necessary living facilities such as a kitchen and toilet can be optimized. For the traditional buildings with relatively general preservation conditions, the renovation and transformation of the interior and exterior of the buildings can be carried out by fully combining the regional characteristics, block positioning and use requirements. The use properties and functions of the building can also be changed without destroying the traditional architectural features. The tourism facilities with distinctive features can be properly developed to meet the needs of Lhasa's new urban development orientation. Traditional buildings that have been poorly preserved can also be rebuilt according to the above-mentioned reuse methods, and the buildings built in violation of the rules can be rebuilt according to the style of traditional buildings. In order to meet the requirement that the new buildings must remain consistent with the traditional style of the entire block, there should be a facade or decoration that is redesigned and reconstructed for the new buildings built in the traditional block that do not conform to its traditional features but have good architectural quality.

4.2 Methods of protective reuse of traditional architecture

For the reuse of traditional buildings in Lhasa, this study mainly discusses the specific methods of reuse of the traditional buildings with relatively general preservation conditions.

4.2.1 Reusing methods based on catering space

Transforming traditional buildings into catering space for reuse is a relatively common method in the protective reuse of traditional buildings, which also has unique advantages in Lhasa. In addition to the traditional Tibetan foods, which have become one of the indispensable experiential aspects of Tibetan tourism, the daily habit of drinking butter tea developed by Tibetans for thousands of years has also resulted in teahouses spreading throughout the streets of Lhasa and it has become an important part of urban and residential life. The "Makye Ame Tea Restaurant" located at the southeast corner of Barkhor Street in Lhasa is an example of one of the numerous renovations of dining space (Fig. 2). This restaurant was opened in a centuries-old two-storey Tibetan-style building in 1997. The facade of the building is yellow and Tibetans also call it the "Yellow House". The first floor's door is red, with traditional Tibetan patterns painted on it. The second floor's window frames are painted black according to traditional customs. The eaves of the doors and windows are adorned with traditional Tibetan decorations. The interior of the building retains the original Tibetan furniture, supplies, utensils and wall decorations, and the newly added furniture and lighting is also based on its coordination with the overall Tibetan style and features (Fig. 3). Makye Ame has become one of the must-see attractions for tourists because of its legends related to the sixth Dalai Lama, Tshangs-dbyangs-rgya-mtsho (Tsangyang Gyatso), and its effective utilization and inheritance of Tibetan culture. However, there are still some inaccurate aspects of Makye Ame. For example, although the yellow exterior wall is one of its most prominent features, based on history and due to the strict religious hierarchy, Tibetan residential buildings' walls are mostly white. Only temples, lingka (parks), palaces and high-ranking religious buildings' walls can be painted with yellow and reddish-brown. In addition, a half open space with a four-angled sloped roof is added to the top floor of the building, which is different from the flat roof of traditional Tibetan dwellings. This space was added to meet the needs of commercial sightseeing. Therefore, in the process of reusing traditional buildings, one of the key points we need to consider is whether the newly added architectural space and decorations need to follow the old social customs and order.

Fig. 2. Outside of makye ame tea restaurant.

Fig. 3 Interior of makye ame tea restaurant.

4.2.2 Reusing methods based on hotels and homestays

The traditional architecture reuse mode based on hotels and homestays can achieve the aim of protective reuse and improve the utilization rate of traditional buildings through the reasonable renovation and repair of traditional residential buildings in Lhasa. In addition, homestays can be used to increase communication and interaction between visitors and locals so that visitors can fully understand the local history while living in a Tibetan cultural environment. There are many mansions and traditional houses in the city that have been transformed into hotels or homestays. Among them, Pandatsang, located at Barkhor South Street, is one of the hostels with many tourists. Founded in 1914, Pandatsang has an area of about 3,000 m^2 and is one of the most well preserved mansions on Barkhor Street. After 1949, it also existed as post offices and courtyards inhabited by more than 70 households. In 2011, Pandatsang was transformed into a youth hostel through investment. The building structure, facade, decoration and roof are preserved so that they are basically intact. The front desk of the hotel is located on the first floor. When it was a mansion, the first floor of the building was usually used for storing goods and raising livestock, while only the second and third floors were used as residential spaces. In addition to the front desk, the colonnade of the first and second floor has been transformed into multi-person rooms for backpackers, and the main building has been converted into twin rooms. There are also problems related to reuse. Traditional Tibetan buildings have smaller spacing between columns. In the process of renovation to hotels, there are limitations in terms of the space separation and functional layout. Furthermore, because Pandatsang's original buildings did not have independent indoor toilets, partition walls had to be built for this purpose. Therefore, when the renovation was completed, Pandatsang covered a large area, but its construction area was small. It has only 40 rooms, and the largest one is 80m^2. (Fig. 4)

Fig. 4 Pandatsang hotel.

4.2.3 Reusing methods based on street-facing shops

Transformation into street-facing shops is one of the most common methods of reuse for traditional buildings in Lhasa. In Barkhor Street, except for the buildings deep in the streets, which are still used as residential space for local residents, the other buildings facing the main streets, especially the first floor, have all been transformed into small shops selling local specialties to tourists. This reuse method gives new functions to traditional buildings that are no longer used for residential space, and promotes the development of the tourism industry in the neighborhood. However, this way of reuse involves a relatively uniform business format and shop content. There is only a single transaction relationship between operators and tourists, and no deeper connection. The renovation of the buildings required an expansion of the openings of the doors and windows on the outer facade of the first floor, and the whole space was adapted for use as a business area. The second or third floors are only used as closed warehouses, so the utilization rate of the buildings is relatively low. The arrival of tourism service-oriented businesses has led to the gradual substitution of tourist-oriented shops for small businesses serving the local people. Tourism service-oriented businesses have become the dominant ones in Lhasa. However, due to the lack of effective control over the quality and type of business, problems that have emerged include insufficient connotations of commercial culture, underestimation of the commercial value of national culture, an excess of counterfeit and inferior commodities, and inadequate protection of commercial heritage. At the same time, merchants and enterprises engaged in tourism business services have become the "new residents" in the neighborhood by renting houses and shops. As a result, most activities in the district are guided by short-term economic interests, which reduces the residents' sense of cultural identification and responsibility. The traditional cultural atmosphere and the continuation of the cultural environment are facing difficulties. To sum up, in terms of the reusing methods based on street-facing shops, we must not blindly follow trends and mechanically copy others. The development of tourism commerce should combine Lhasa's inherent commercial culture, folk culture and various handicraft techniques to establish a benign protection mechanism for intangible cultural heritage, build traditional handicraft workshop areas, concentrate the local handicraft industry in the surrounding area, promote the productive protection of handicraft skills, and promote the revival of "old-fashioned" commerce. (Fig. 5)

Fig. 5 Shops on barkhor street.

5 Summary

As an indispensable and characteristic part of the Chinese architecture system, traditional Tibetan architecture in Lhasa has a great value in relation to protection and inheritance. Compared with the single static protection mode, the dynamic reuse of buildings is a more effective means for activating architectural cultural heritage. Protective reuse emphasizes reasonable restoration, adapts to local conditions, and must not blindly follow trends and transform traditional buildings in a way that causes them to lose their regional and cultural characteristics. The protection and reuse of traditional buildings in Lhasa should also be fully combined with the orientation of Lhasa's future development. In order to promote the sustainable development of Lhasa's history and culture, and under the premise of tourism utilization, the protection of architectural heritage should be changed from simply improving living conditions to establishing the cultural identity of new and indigenous residents, as well as stimulating the enthusiasm of residents to participate in the protection of cultural heritage and encouraging them to actively pass down traditional living customs.

Acknowledgments

This work was supported by the National Key R&D Program of China (2016YFC0700400). The authors thank the reviewers for their helpful comments and suggestions.

References

1. Z Xu. An introduction to Tibetan traditional architecture, first ed. China Architecture & Building Press (2004)
2. S Yu. An overview of the development of Tibet architecture and decoration. Architectural Journal. **6** (2013)
3. L Huang, C Liu. The image relationship between traditional architectural space and religious culture in Tibet. Huazhong Architecture. **28**, 5 (2010)
4. Y Peng, M Cao. Freely Developed Traditional Buildings in Tibet——Primary Research on the Planning and Space Features in Tibet Traditional Buildings. Chinese and Overseas Architecture. **10** (2008)
5. J Zhuang. Traditional houses in Lhasa and architectural styles of Tibet. Planners. **2** (1995)
6. D Xu, X Dong. The social function of Tibetan folk. Journal of Sichuan University for Nationalities. **19**, 1 (2010)
7. Y Chen, S Liu. The influence of Tibetan Buddhist color culture on Tibet residential architecture. Pubic Communication of Science & Technology. **2** (2015)
8. J Zhou, T Li. Study on the Origins of Tibetan Rdzong(fort)——the Official Building. Journal of Tibet University. **23**, 5 (2008)

Properties and Characteristics of Compressed Biofuel Production from Empty Palm Fruit Bunches

Tanuwat Larptansuphaphol* *and Penja* Jitjumroonchokchai

Department of Biotechnology, Faculty of Applied Science, King Mongkut's University of Technology North Bangkok, Thailand.

Abstract. Thailand sits in the third place of the global rankings for crude palm oil production with an annual output of approximately 2 million tons/p.a., or 1.2% of the global output. Empty palm fruit bunches is one of residual from palm oil extraction process of Palm Oil industry. The objective of this research was to study properties and characteristics of biofuel made from empty fruit bunches and mixed with black rice hush with various composition ratios of empty palm fruit bunches and black rice husk as 100:0, 90:10, 80:20, 70:30, 60:40 and 50:50. Only the biofuels with a ratio of 100:0 and 90:10 could be formed to be compressed biofuels. The heating values of biofuels were analyzed by Gallenkamp Autobomb Calorimeter whereas other fuel properties and characteristics were tested and analyzed according to ASTM standards. The comparison results between biofuel from a mixture ratio 100:0 and 90:10 showed that the average of heating value, ash content, carbon content and combustion time (minute per one piece) were 4,672.46 and 4,199.33 cal/g, 21.73% and 29.30%, 12.07% and 16.28%, 92.33 and 63 minutes, respectively.

1. Introduction

The Oil Palm plant (Elais Guinensis) produces a fruit with a nut inside the fruit. The fruit is boiled in water to extract the palm oil. After the oil extraction from the fruit, the nuts are broken for the palm kernels. The broken shells are called palm kernel shell. In the palm oil processing operations, the solid wastes are: empty fruit bunches, palm fibre, and palm kernel shell. In rural community levels, palm kernel shells are used in their crude form for heating during cooking. However it burns with a lot of smoke due to its organic content. The effect of such smoke to health cannot be over emphasized. Even in such cases most of the energy content is not used up as palm kernel shell charcoals which are common sight at ash dumps [1].

Palm empty fruit bunches is waste residue generated from palm oil industries. After harvesting fresh fruit bunches from oil palm tree, these bunches are sterilized in a horizontal steam sterilizer to inactivate enzymes present in pericarp and loosen fruits from bunches. The sterilized bunches are fed into a rotary drum thresher in order to remove the sterilized fruit from bunches.

These bunches without fruit are called as empty fruit bunch which are conveyed to the damping ground, whereas the sterilized fruits are further used as feedstock for palm oil production in palm oil extraction process by means of screw type press. The effluents from screw type press are nuts and fibers which are separated from each other by cyclone. After this separation, nuts are cracked into shells and kernels [2].

The palm oil plants are found in most parts, especially, southern states of Thailand. The different parts of the oil palm are adapted for different useful purposes. While the leaves provide brooms for tidying the environment, the kernel is a major source of red palm oil used for cooking. The Palm empty fruit bunches were converted into briquettes by adopting very simple process, utilizing readily available and affordable materials [3]. The briquettes produced from Palm empty fruit bunches were tested for their combustion and physical characteristics. Summers studied about using rice straw for energy production and told that rice straw has a higher value about 16,000 MJ/Kg and meaning a potential fuel base of 1.8x1010 MJ in the rice growing region of California [4]. Chou et al. investigated the feasibility of preparing the biomass briquettes by the solid waste, such as rice straw and rice barn [5]. Biomass pellets are generally a superior fuel when compared to their raw feedstock. Not only are the pellets more energy dense, they are also easier to handle and use in automated feed systems. These advantages, when combined with the sustainable and ecologically sound properties of the fuel [6]. Global waste biomass energy capacity is about eight times the annual world consumption of energy from all sources. Therefore biomass represents a very large energy resource, in Eastern Europe was used biomass energy in industrial more than 10 years and this low

* Corresponding author: tanuwat.l@sci.kmutnb.ac.th

burning rate restricts the use of these biomass energy to small capacity applications such as stoves and domestic heating furnaces [7, 8]. .

This project of bio-fuel made from Palm empty fruit bunches aimed to reduce greenhouse gases such as carbon dioxide (CO2) from biomass waste in open burning areas. The objective was to make bio- fuels as a renewable energy source. Utilization wastes from palm oil industry were cover processing of bio-fuels, costs and environmental benefits.

2. Material and Methods

Palm empty fruit bunches were collected from Thai Oil Industry Co., Ltd. at Samutsakorn province. Black rice husk was added in a ratio of palm empty fruit bunches: black rice husk (by weight) as 100:0: , 90:10, 80:20, 70:30, 60:40 and 50:50 respectively. Bio-fuel was compressed by using screw machine to form a 6 cm. diameter and 10 cm. height cylinder shape fuel. Palm empty fruit bunches as shown below in Figure 1.

Fig. 1 Dried palm empty fruit bunches

Properties and characteristics of compressed bio-fuel made from palm empty fruit bunches were tested and analyzed as follows: moisture content, volatile components, ash, carbon, hydrogen, oxygen, nitrogen and sulfur under ASTM (American Testing Standard Materials) D3173, D3174 and D4239. Gross calorific value was measured by using Gallenkamp Autobomb Calorimeter Instrument according to ASTM D5685. Experimental data were analyzed by analysis of variance (ANOVA) and Duncan's New Multiple Range (DMRT) at significant 5%. The schematic diagram of statistical analysis as shown below:

T= Ratio of empty fruit bunches: black rice husk

T1 =100:0, T2 =90:10, T3 = 80:20, T4 = 70:30, T5 = 60:40 and T6=50:50

R= sample

R1 = sample1, R2 =sample3 and R3 = sample3

T1R1	T2R1	T3R1	T4R1	T5R1	T6R1
T1R2	T2R2	T3R2	T4R2	T5R2	T6R2
T1R3	T2R3	T3R3	T4R3	T5R3	T6R3

3. Results and Discussion

3.1 Physical characteristics

General physical characteristics of compressed fuel made from Palm empty fruit bunches that were taken from Thai Oil Industry Co., Ltd. at Samutsakorn province were tested at KMUTNB laboratory. The physical biomass fuels were shown in Figure 2. The physical and chemical characteristics of compressed bio-fuel made from Palm empty fruit bunches were shown in table1.

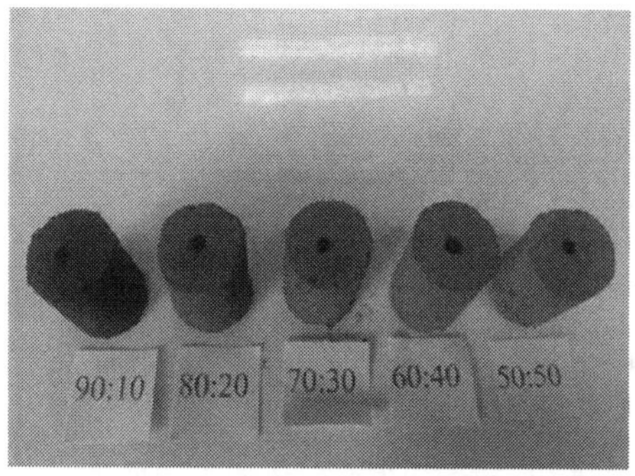

Fig 2. Physical characteristics of compressed bio-fuel made from Palm empty fruit bunches

Table 1 Physical characteristics of compressed fuel made from Palm empty fruit bunches

Characteristics	% by weight of Palm empty fruit bunches (one piece)
Moisture content (%)	18.22
Ash (%)	6.55
Fixed carbon (%)	38.10
Bulk Density, kg/m^3	288.05

3.2 Physical and chemical properties of compressed bio-fuel

After compressed bio-fuel was dried in an open area for 7 days, all samples of 6 ratios of Palm empty fruit bunches: black rice husk were tested and analyzed in a laboratory. The samples were tested and analyzed for their physical and chemical properties of biomass energy according to ASTM standard methods. The results of testing and analysis of compressed bio-fuel made from palm empty fruit bunches: black rick husk were shown in Table 2.

Table 2 Physical and chemical properties of compressed bio-fuel at various ratios

Physical/Chemical properties	100:0	90:10	80:20	70:30	60:40	50:50
Moisture content (%)	17	20	18	17	21	16
Volatile matters (%)	87	86	88	85	84	83
Ash content (%)	13	17	19	20	20	16
Fixed carbon (%)	30	33	35	34	35	37
Hydrogen (%)	4.1	4.8	4.8	4.9	4.0	3.9
Oxygen (%)	41	44	46	40	42	45
Nitrogen (%)	0.55	0.48	0.53	0.49	0.39	0.33
Sulphur (%)	0.19	0.17	0.18	0.21	0.23	0.18

Table 2 showed the results of proximate analysis (volatile matters, fixed carbon and ash content) calculating on a dry and ash-free basis. The ultimate analysis was analyzed for all elements such as carbon(C), hydrogen(H), Nitrogen(N), Oxygen(O) and Sulphur(S) by using Gas Chromatography technique. The results indicated that ratio of palm kernel shell: black rice husk as 50:50 had the lowest moisture content 16%.

European Biomass Association suggested that a moisture content of raw materials to be a compressed fuel should be 12-17% and it was not exceed these values [9]. In addition, compressed fuel should be stored in a closed system or a room, a silo, bunkers, plastics and air light zip bags (dry conditions) and storage times should be minimized to prevent absorption from atmospheric moisture. However, the bulk material supply was necessary and available in all seasons [10].

3.3 Heating value

Higher heating value was experimentally determined in a bomb calorimeter by concealing a stoichiometric mixture of fuel and oxidizer (e.g., two moles of hydrogen and one mole of oxygen) in a steel container at 25 °C. Fuel sample was initiated by an ignition device and the combustion reactions completed. When hydrogen and oxygen reacted during combustion, water vapor emerged. Subsequently, a vessel and its content were cooled down to an original temperature 25 °C and the higher heating value was determined as the heat released between identical initial and final temperatures [11] .

The CV (Calorific Value) of a substance was measured by burning it in a controlled environment. The result of heat released by this combustion i.e. the net temperature rise, was proportional to the calorific value. In an adiabatic system the environment was controlled so that no energy lost or gained. In order to achieve this state, a Calorimeter Vessel (bomb) and a bucket were surrounded by a water jacket which was strictly temperature controlled by a system of circulators, heaters and coolers, so that it had the same temperature as an inner bucket. The Isothermal calorimeter measured an effect that the environment had on the bucket before and after determination and correction of the results accordingly. This implied that the environment was stable during the determination. In order to achieve a stability condition, a large body of water surrounding the vessel and bucket assembly was thermally stable [12].

Heat of combustion was transferred to cooling water around Bomb calorimeter. Water temperature was measured by thermometer. Amount of heat transfer was calculated by equation (1).

$$Q = mc\,(T_1 - T_2) \tag{1}$$

where

Q = fuel combustion heat
m = mass of water in Calorimeter
c = specific heat of water
T_1 = water temperature before combustion
T_2 = water temperature after combustion

Fuel thermal energy could be determined from equation (2).

$$HHV = \frac{(\Delta T \cdot w) - e}{g} \tag{2}$$

where

HHV = Combustion heat
ΔT = heat increasing temperature (°C)
W = 1,724.187 (cal/°C)

e = combustion heat correction factor = 2.3 (cal/cm) x combustion wire length (cm)
g = fuel sample weight (g)

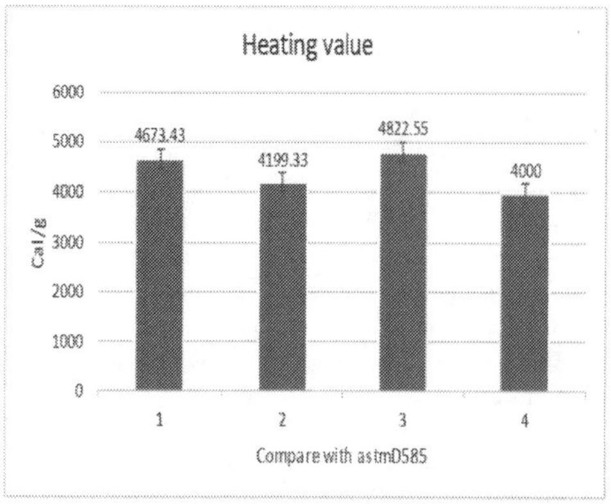

1/ Palm empty fruit bunches 100%
2/ Palm empty fruit bunches : black rice husk 90:10
3/ Wood compressed bio-fuel
4/ Standard ASTM D5865

Fig. 3 Comparison of heating value of various fuels with ASTM D5865

Figure 3 showed the comparison results of heating values of various samples with standard ASTM D5865. The results indicated that the heating value by Bomb Calorimeter was about 4,672.43 cal/g and 4,199.33 cal/g from the ratio of Palm empty fruit bunches: black rice husk at 100:0 and 90:10 respectively.

3.4 Determination of heat utilization efficiency

Thermal behavior of compressed bio-fuel had been studied by measuring the rate of weight loss of the samples as a function of time and temperature. Thermal combustion energy could be measured by the time from initial to final until water was boiled at its boiling point. The maximum average time for heat utilization efficiency was about 92.33 minutes from the ratio of Palm empty fruit bunches 100% no addition of black rice husk.

Various ratios of combiner mixtures were analyzed for the best combination of mixtures. The ratio of Palm empty fruit bunches: black rice husk palm as 90:10 was the best combiner. The highest temperature of this compressed fuel at 63 minutes of combustion time at 99.31 o C. Percentage of ash from combustion in 60 minutes.

Ash content was determined by percentage of compressed fuel combustion depending on the time of experiment in 60 minutes. The lowest ash content was the result at ratio of Palm empty fruit bunches: black rice husk as 90:10. From the results showed that ash contents of 6 ratios of samples after 60 minute that 5 ratios of samples mixed with binder had higher ash content compared with compressed wood fuel. Ash content in compressed wood fuel was usually less than 1%, while it could be very high in many agricultural residues. That was why compressed wood fuel quality standards could not be reached when using agricultural residues as feedstock and another, specified quality standard for agricultural compressed fuel needed to be introduced [13].

4 Conclusion

The advantages of compressed fuel production are as follows: alternative energy for people both in the urban and rural areas, wastes reduction by agricultural wastes reuse and recycle, renewable energy from living organisms, alternative fuel market for low cost. Moreover, compressed fuel made from Palm empty fruit bunches can be developed for export to increase national incomes.

The rise in energy costs as interlinked of food prices. The residues from agriculture area used to be renewable energy is the way to help people in developing country to use cheap energy and reduce growing plants for energy. That means food and energy security is the safety way of human being in the future.

The percentage of ash content measured from Bio-fuel still higher than wood (1-2% of ash). In addition, ash content can originate from the biomass itself or from binder and often lead to high disposal costs resulting in a negative economic impact on biomass conversion systems. Ash content values may vary significantly from one biomass resource to another. In this energy conversion processes it is not only the amount of ash, but also its physical and chemical properties that should be considered.

Finally, the government should start looking into providing the necessary infrastructure, technology and expertise for the use of potential power supply from oil mills and other biomasses (like solid wastes).
Furthermore, the private sector such as SMEs should be a partner with the government for funding and investment for energy stability in the country.

The author would like to thank for Department of Mechanical Engineering, Faculty of Engineering, King Mongkut's University of Technology, North Bangkok for analytical support of heating values by Bomb Calorimeter Instrument.

References

1. UGWU,KE;AGBO, KE.Briquetting of Palm Kernel Shell. J. Appl. Sci. Environ. Manage. Sept, Vol. 15 (3) 447 - 450(2011).

2. Abd, Halm Shamsuddin. Palm Oil Products and Wastes as Alternative Energy Sources. PERIC International Energy Journal: Vol 11, No.2, Dec. 1989.
3. Obuka Nnaemeka, Onyechi Pius, Okoli Ndubuisi. Palm Oil Biomass Waste a Renewable Energy Resource for Power Generation Saudi J. Eng. Technol., Vol 3, Iss-12 (Dec, 2018): 680-691.
4. Matthew D. Summers .Using Rice Straw for Energy Production: Economics, energetics and Emissions. http://jenkins.ucdavis.edu/projects/RiceStraw/RiceStrawDocs/SummersEBS216FinalReport.pdf(2010).
5. C.S Chou, S.H. Lin, W.C. Lu, Preparation of solid biomass fuel made from rice straw and rice bran, Fuel Processing Technology (9), pp. 980-987. DOI :10.1016/j/fuproc_2009.04.012. (2009).
6. Manufacturing Fuel Pellets from Biomass. www.pubs.cas.psu.edu/Free Pubs/pdf/uc202 pdf.(2010)
7. Berger H. Zimmer H J and Garmo J. Ontersuchngen zur charakateristerung von strohpellets G. verdaulich- keitsuntersuchungen an weizenstrohpellets. Archiv. Tiermahr. 24, pp. 689-700. (1984)
8. Ewida K.T., H EL-Salmawy, N.N. Atta and M.M. Mahmoud . A sustainable approach to the recycling of rice straw through pelletization and controlled burning. Clean Tech environ Policy Journal, 8, pp. 188-197. DOI 10.1007/s10098-006-0043-x. (2011)
9. European Biomass Association (2000). http://www.energyagency.at/(en)/publ.(2010).
10. Bergram PCA, Kiel J. A. . Torre faction for biomass upgrading published at 14th European Biomass Conference and Exhibition, Paris, France, 17-21 ECNRX-05-180, 2005-12-18.(2005).
11. http://en.wikipedia.org/wiki/Heat_of_combustion (2011).
12. http://www.cal2k.com/article_10.asp(2013).
13. Significant biomass properties. Part 2: Ash content. http://www.biomassenergy.gr/en/articles/biomass/315-significant-properties-part-2-ash-content(2014).

Characterization of Torrefied Biomass from Sugarcane (*Saccharum officinarum*) Bagasse Blended with Semirara Coal

Rose Ann P. Lomeda-De Mesa[1,4], *Allan N.* Soriano[3], *Ariziel Ruth D.* Marquez[1,2], and *Adonis P.* Adornado[1,2,4,*]

[1]School of Graduate Studies, Mapúa University, Manila 1002, Philippines
[2]School of Chemical, Biological, and Materials Engineering and Sciences, Mapúa University, Manila 1002, Philippines
[3]Department of Chemical Engineering, Gokongwei College of Engineering, De La Salle University, 2401 Taft Avenue, Manila 1004, Philippines
[4]General Education Department, Colegio de Muntinlupa, Mayor J. Posadas Avenue, Sucat 1770, Muntinlupa City, Metro Manila, Philippines

Abstract. Utilizing and improving the quality of Philippine indigenous coal is getting more important in order to sustain the country's high dependence on coal over the next 10 years and to keep up with the country's growing energy demands. Also, environmental problems and negative impacts of agricultural wastes are drawing more and more attention since the quantity of agricultural wastes has been rising rapidly all over the world. In the Philippines, sugarcane (*Saccharum officinarum*) bagasse has been identified as one of the significant contributor to the country's biomass energy resource. Hence, in this study, coal blending technique has adopted as a proper approach to improve the quality of indigenous coal reserves while concurrently reducing and reusing agricultural wastes. This paper aimed to establish recommended blending ratios for Semirara coal, a sub-bituminous type of coal from Semirara Island in the province of Antique, Philippines and sugarcane (*S. officinarum*) bagasse. Proximate analysis, ultimate analysis, and calorific value were determined to characterize and understand the physical conditions and coal properties during combustion. Results showed that blending torrefied sugarcane (*S. officinarum*) bagasse with Semirara coal would generally improve its quality in terms of its combustion properties thereby making these combinations of coal and biomass advantageous.

1 Introduction

Coal is the largest source of energy in power generation industry worldwide that supports for the economy in both developed and developing countries. Currently, coal fuels 46% of the world's electricity and foreseen to continue to supply a strategic share over the next three decades [1]. The Asian Development Bank (ADB) said that the Philippines is expected to become the most coal-dependent country in Southeast Asia by 2030 [2]. In the Philippines, one coal resources is in Semirara Island located in Caluya, Antique, with an estimated total coal reserve of around 170 million metric tons [3]. However, Semirara coals are black in color, dull (not shiny), have higher moisture and volatile matter, and lower sulfur and fixed carbon content compared to other types of coal, referred to as sub-bituminous [4,5]. Moreover, it is classified as second lowest rank of coals due to their lower calorific value [6-8].

Considering that the Philippines is also an agricultural country wherein sugarcane (*Saccharum officinarum*) is the number one agricultural crop produced in the country in terms of tonnage, generating an enormous abundance of agricultural biomass (bagasse) resource of approximately 6 million metric tons yearly based on the data of Bureau of Agricultural Statistics, that may be used as energy fuel [9].

Utilizing and improving the quality of Philippine indigenous coal is getting more important in order to sustain the country's high dependence on coal and to ensure that the growing energy demand can also be addressed. Likewise, in recent years, the quantity of agricultural waste has been rising rapidly all over the world. As a result, the environmental problems and negative impacts of agricultural waste are drawing more and more attention. Therefore, there is a need to adopt proper approaches to reduce and reuse agricultural waste.

In order to simultaneous address these issues, one of the cost-effective methods to develop the indigenous coal reserves is via coal blending with biomass, essentially blending two dissimilar fuels, with one being the dominant fuel and the other its supplement. Typically coal is the dominant fuel, and the principles of blending apply, thereby making some combinations of biomass and coal advantageous. Through this technique, it ensures that the cleaned coal contains minimum ash, sulfur, and also maximum heating value on a timely basis.

Thus, the main purpose of this work is to establish recommended blending ratios for sugarcane (*S. officinarum*) bagasse and Semirara coal. Proximate

* Corresponding author: apadornado@mapua.edu.ph

analysis, ultimate analysis, and calorific value were determined to characterize and understand the physical conditions and coal properties during combustion process.

Furthermore, this study would give a valuable introduction on the combustion characterization of Philippine coal and its blends that will enrich the database on properties of Philippine coal and of Philippine coal blends, and would show the potentials and fuel flexibility of the Philippine coal.

2 Methodology

2.1 Preparation of Sugarcane (*Saccharum officinarum*) Bagasse

For this study, 3 kilograms of raw sugarcane (*Saccharum officinarum*) bagasse were collected from a sugarcane plantation in the Town of Calaca in the province of Batangas, Philippines. Representative sugarcane (*S. officinarum*) samples were taken to the Bureau of Plant Industry (BPI) for plant identification, species authentication, and certification.

The raw sugarcane (*S. officinarum*) bagasse was reduced in size in the range of 1.0 to 1.5 cm, and then dried using a Binder Drying Oven ED 53 at 105°C for 24 hours prior torrefaction. The weights of the raw and the oven-dried sugarcane (*S. officinarum*) bagasse were recorded, and then the percentage of moisture removed from drying was calculated.

2.2 Torrefaction of Sugarcane (*Saccharum officinarum*) Bagasse

Biomass torrefaction is a mild thermochemical process that generally performed in inert atmosphere in the temperature range of 200 to 300°C for several minutes or hours. This treatment gained interest as it promotes an increase in the energy content of biomass to levels equal to and sometimes above that of coal [10], making it a more attractive and competitive source within the primary energy matrix [11].

The oven-dried sugarcane (*Saccharum officinarum*) bagasse was torrefied using an electric Ney Vulcan 3-1750 Box Muffle Furnace equipped with thermocouple, slide out heating elements, combined fiber, and firebrick insulation for quick cooling and three-stage programmable system of temperature regulation. The 1,750 cubic inches inner volume capacity of the furnace was connected to the outer atmosphere through a tube that provides access of oxygen into the torrefaction zone and discharge of gaseous pyrolysis products.

For each run, approximately 200 grams of oven-dried sugarcane (*S. officinarum*) bagasse were placed into unsealed 110.45 cubic inches ceramic vessels to provide free access of oxygen. The oven-dried sugarcane (*S. officinarum*) bagasse was torrified at 300°C (heating rate of 5°C/min) for 30 minutes under normal atmospheric pressure. The torrefied sugarcane (*S. officinarum*) bagasse was grinded using a Wiley mill and then sifted using a stainless steel laboratory test sieve with mesh screen no. 60 equivalent to an approximate size of 250 µm to homogenize the particle size. Finally, the sifted sugarcane (*S. officinarum*) bagasse were collected in a clean, dry, and sealed bag, and stored in a dessicator until used for blending and characterization.

The weights of the oven-dried and the torrified sugarcane (*S. officinarum*) bagasse were recorded, and then the percentage recovery was calculated.

2.3 Proximate Analysis

The proximate analysis performed for the torrefied sugarcane (*Saccharum officinarum*) bagasse, Semirara coal, and its blends (% w/w) – 25, 50, and 75 was in accordance to ASTM D3172-13: Standard Practice for Proximate Analysis of Coal and Coke [12]. This practice covers the determination of moisture (MC), volatile matter (VM), and ash (AC) and the calculation of fixed carbon (FC).

All measurements performed in this study were carried out in three replicate runs and the mean values were reported.

2.4 Ultimate Analysis

The ultimate analysis for the torrefied sugarcane (*Saccharum officinarum*) bagasse, Semirara coal, and its blends (% w/w) – 25, 50, and 75 was performed using LECO CHN628 Series Elemental Determinator and CKIC 5E-S3200 Coulomb Sulfur Analyzer operated using S3200 application program to determine the carbon (C) / hydrogen (H) / nitrogen (N) and the total sulfur (S) content, respectively.

For C/H/N determination, a 0.10 mg of the torrified sugarcane (*S. officinarum*) bagasse was placed in a tin capsule, heated at 980°C with a constant flow of helium enriched with oxygen gas [13].

While for the total S, on average, a 50.45 mg of the torrified sugarcane (*S. officinarum*) bagasse is required, and then covered with a thin layer of tungsten trioxide (WO_3) after weighing in temperature rise period. The furnace was heated to 1,150°C for 30 minutes before analyzing the samples.

Also, average values of the triplicate runs were reported for all measurements done in this study.

2.5 Calorific Value (Experimental)

The experimental calorific value (CV) of the torrefied sugarcane (*Saccharum officinarum*) bagasse, Semirara coal, and its blends (% w/w) – 25, 50, and 75 was determined using an adiabatic CKIC 5E-AC/PL Calorimeter by completely combusting 1.00 gram of the samples under a pressurized O_2 atmosphere [10]. For all combustion experiments, the samples were charged with oxygen at a pressure of 2.80 to 3.00 MPa and a cylinder pressure of ≥ 5 MPa.

2.6 Calorific Value (Calculated)

The values obtained from the proximate and ultimate analyses were used to calculate the calorific value based on existing correlations taken from the literature which

 https://doi.org/10.1051/e3sconf/201912002002

are applicable for both coal and biomass based on parameters for proximate and ultimate analyses as shown in Tables 1 and 2, respectively. Thus, it follows that the existing correlations presented are not specific for sugarcane (*Saccharum officinarum*) bagasse alone.

Table 1. Correlations for calorific value applicable for both coal and biomass based on proximate analysis.

Study	Correlation for Calorific Value, MJ·kg^{-1}	Eq.
Kavšek et al. [14]	CV = 0.4108FC + 0.1934VM - 0.021AC	(1)
Kieseler et al. [15]	CV = 0.3536FC + 0.1559VM - 0.0078AC	(2)
Majumder et al. [16]	CV = -0.03AC - 0.11MC + 0.33VM + 0.35FC	(3)
Mesroghli et al. [17]	CV = 37.777 - 0.647MC - 0.387AC - 0.089VM	(4)
Parikh et al. [18]	CV = 0.3536FC + 0.1559VM - 0.0078AC	(5)

Table 2. Correlations for calorific value applicable for both coal and biomass based on ultimate analysis.

Study	Correlation for Calorific Value, MJ·kg^{-1}	Eq.
Channiwala & Parihk [19]	CV = 0.3491C + 1.1783H + 0.1005S - 0.1034O - 0.0151N - 0.0211A	(6)
Friedl et al. [20]	CV = 0.00355C^2 - 0.232C - 2.23H + 0.0512CH + 0.131N + 20.6	(7)
Jenkins et al. [21]	CV = -0.763 + 0.301C + 0.525H + 0.064O	(8)
Sheng & Azevedo et al. [22][b]	CV = 0.3259C + 3.4597	(9)
Sheng & Azevedo et al. [22][c]	CV = -1.3675 + 0.3137C + 0.7009H + 0.0318O^{*a}	(10)
Tillman [23]	CV = 0.4373C -1.6701	(11)

[a] O* is the sum of the contents of oxygen and other elements in the organic matter (O* = 100-C-H-Ash).

3 Results and Discussion

The results for the proximate analysis, ultimate analysis, and calorific value for the torrefied sugarcane (*Saccharum officinarum*) bagasse, Semirara coal, and its blends (% w/w) – 25, 50, and 75 were summarized in Tables 3, 4, and 5, respectively.

Table 3. Results for the proximate analysis – moisture (MC), ash (AC), volatile matter (VM), and fixed carbon (FC) on the basis of concentration (% w/w) of sugarcane (*Saccharum officinarum*) bagasse.

Concentration (% w/w) of Biomass	MC (%)	AC (%)	VM (%)	FC (%)
0	13.04	36.03	24.19	26.74
25	9.64	39.04	29.20	22.12
50	7.36	40.86	33.13	18.65
75	5.64	43.73	36.67	13.96
100	2.89	45.24	42.22	9.65

Table 4. Results for the ultimate analysis – carbon (C), hydrogen (H), nitrogen (N), and sulfur (S) on the basis of concentration (% w/w) of sugarcane (*Saccharum officinarum*) bagasse.

Concentration (% w/w) of Biomass	C (%)	H (%)	N (%)	S (%)
0	40.22	4.61	9.80	0.77
25	45.57	4.17	9.85	0.80
50	52.00	3.81	9.66	0.60
75	57.08	3.53	9.65	0.56
100	61.48	3.19	9.78	0.38

Table 5. Results for calorific value and energy yield on the basis of concentration (% w/w) of sugarcane (*Saccharum officinarum*) bagasse.

Concentration (% w/w) of Biomass	Calorific Value (MJ·kg^{-1})	Energy Yield (%)
0	16.17	-
25	18.57	42.68
50	20.16	46.33
75	21.82	50.15
100	23.20	53.32

3.1 Proximate Analysis

Generally, any two coals cannot just be blended. To decide to blend or not, it is very important to understand the composition of the coals that are to be blended. The linear additive rule shown in Equation (12) is used to estimate the theoretical composite value of the coal blends to show the relationship of the coal properties and the amount of that coal in the blend, where M is the composite value of the parameter being estimated and x is the weight fraction of the components in the blend.

$$M = x_b M_b + (1 - x_b) M_a \quad (12)$$

A property can be considered additive when the physical property of the blend can be predicted by the relative amounts of component coals and their physical properties. The applicability of the additive property for coal blending was determined using the coefficient of determination, R^2 value of at least 0.9000 [24]. Based on Figure 1, comparing the experimental values from the calculated values for moisture (MC), ash (AC), volatile matter (VM), and fixed carbon (FC) based on the linear additive rule, it showed acceptable results with minimal deviations with overall R^2 value of 0.9602. Thus, it can be established that Semirara coal blended with sugarcane (*Saccharum officinarum*) bagasse were additive based on its proximate analysis parameters.

The proximate analysis involves the quantitative determination of moisture content (MC), ash content (AC), volatile matter (VM), and fixed carbon (FC). The results for the proximate analysis for the torrefied sugarcane (*S. officinarum*) bagasse, Semirara coal, and its blends (% w/w) – 25, 50, and 75 was shown in Table 3. The trends of these values were also illustrated in Figure 1 on the basis of increasing concentration (% w/w) of sugarcane (*S. officinarum*) bagasse.

 https://doi.org/10.1051/e3sconf/201912002002

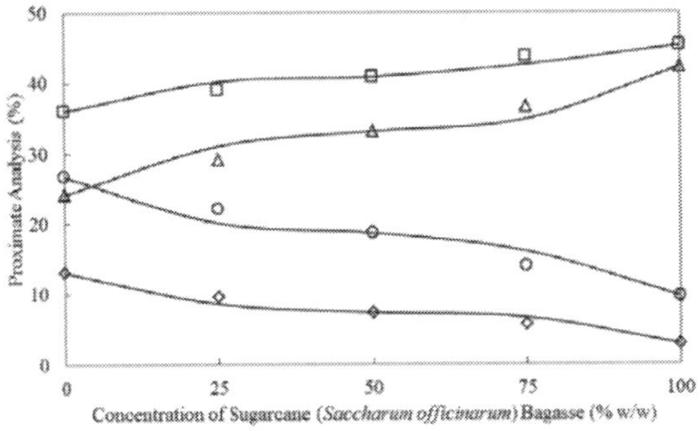

Fig. 1. Comparison of experimental values for proximate analysis parameters – moisture (◇), ash (□), volatile matter (△), and fixed carbon (○) versus calculated values (solid lines) based on linear additive rule.

It was observed that an increase in the concentration of torrefied sugarcane (*S. officinarum*) bagasse in the coal blend caused a decrease in the moisture content (MC) by as much as 77.84%. An increase in the moisture content of coal results in an increase in heat loss due to evaporation and superheating of vapor. Therefore, an increase in the concentration of torrefied sugarcane (*S. officinarum*) bagasse in the coal blend is a good indication that it averts the presence of fungi and decreases other biological activities upon storing. High moisture content can affect the overall energy conversion upon combustion because it influences the calorific value of the given fuel. The higher moisture content, the less energy per kilogram is released that causes lower calorific value [25]. This claim could be supported by the relationship of moisture content and calorific value shown in Table 5. A decrease in the moisture content in the coal blend corresponds to an increase in the calorific value.

Ash (AC) is the inorganic residue after combustion. One characteristic of a good fuel is having little amount of ash during combustion. Small amount of ash residues signifies the maximum utilization of the fuel used. High amount of ash affects combustion efficiency and boiler efficiency because it causes clinkering and slagging. Looking at Table 3, it can be observed that an increase in the amount of torrefied sugarcane (*S. officinarum*) bagasse in the coal blend also increases the amount of ash.

High volatile matter (VM) indicates that the fuel can be easily ignited and subsequently oxidized. High volatile matter also contributes to better efficiency for burning during combustion. It influences the secondary air requirement, distribution aspects, and secondary oil supports. An increase in the amount of torrefied sugarcane (*S. officinarum*) bagasse in the coal blend corresponds to an increase in volatile matter by 74.54%.

Fixed carbon (FC) is the solid combustible residue that remains after a biomass particle is heated and the volatile matter is expelled. The amount of fixed carbon and volatile combustible matter directly contribute to the heating value of coal. Fixed carbon acts as a main heat generator during burning. High fixed carbon in biomass increases the char formation [25]. An increase in the amount of torrefied sugarcane (*S. officinarum*) bagasse in the coal blend caused a decrease in the amount of fixed carbon by 63.91%.

Lastly, in terms of the coal blending ratios used in this study, the 75 % w/w Semirara coal blended with 25 % w/w torrefied sugarcane (*S. officinarum*) bagasse is considered to be the most favorable based on the results of the proximate analysis.

3.2 Ultimate Analysis

Shown in Figure 2 is the comparison of experimental values versus the calculated values based on the linear additive rule for ultimate analysis – carbon (C), hydrogen (H), nitrogen (N), and sulfur (S). Just like in proximate analysis, the values are within the allowable results with slight deviations with overall R^2 value of 0.9349. Thus, it follows that Semirara coal blended with sugarcane (*Saccharum officinarum*) bagasse were also additive based on its ultimate analysis parameters.

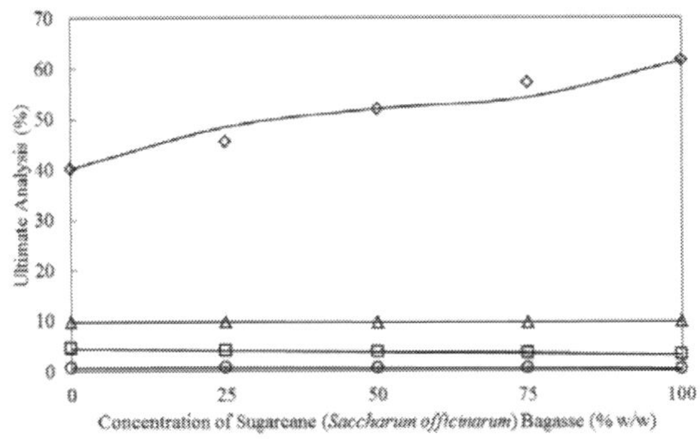

Fig. 2. Comparison of experimental values for ultimate analysis parameters – carbon (◇), hydrogen (□), nitrogen (△), and sulfur (○) versus calculated values (solid lines) based on linear additive rule.

The ultimate analysis determines the elemental composition of the coal including moisture, ash, carbon (C), hydrogen (H), nitrogen (N), sulfur (S), and oxygen (O) (by difference). It is valuable in determining the quality of air required for combustion and composition of combustion gases. The results for the ultimate analysis for the torrefied sugarcane (*S. officinarum*) bagasse, Semirara coal, and its blends (% w/w) – 25, 50, and 75 was shown in Table 4. The trends of these values were also illustrated in Figure 2 on the basis of increasing concentration (% w/w) of sugarcane (*S. officinarum*) bagasse.

Carbon content is a good indicative property of coal ranks. The highest rank coals have the highest carbon contents and the lowest oxygen contents. Ratios of these elements can indicate the rank of the coal and its coalification degree [26]. It is also important to measure the sulfur content in coal samples to evaluate the potential sulfur emissions from coal combustion. Increasing the amount of torrefied sugarcane (*S. officinarum*) bagasse in the coal blend increases the carbon content by 52.86%.

Meanwhile, the hydrogen content for all coal blends decreases by 45.12% with increasing amount of torrefied sugarcane (*S. officinarum*) bagasse. The increased content of hydrogen is normally more of a characteristic of low rank coals, while the decreased values are commonly more typical for higher-rank coals [27]. Based on this, it could be said that a decrease in hydrogen content indicates that the quality of Semirara coal blended torrefied sugarcane (*S. officinarum*) bagasse improved to higher rank.

Nitrogen content for all coal blends is approximately the same with 9.75% on average. Nitrogen is an inert and incombustible gas and it does not contribute any useful property to the combustion of coal. It is generally found in the organic fraction of the coal wherein upon combustion, it is emitted as nitrogen oxide (NO_x) in the flue gas.

Sulfur content for all coal blends decreases by 59.74% with increasing amount of torrefied sugarcane (*S. officinarum*) bagasse. Presence of sulfur in coal is harmful for use in metallurgy as it transfers to the metal and adversely affects the property of metal. Also, oxidation products of sulfur such as sulfur dioxide (SO_2) and sulfur trioxide (SO_3) have corrosive effect on the equipment and causes atmospheric pollution. Hence the presence of sulfur in coal is undesirable. A decrease in the sulfur content of Semirara coal upon coal blending with torrefied sugarcane (*S. officinarum*) bagasse is an indication that an environment-friendly approach have been taken.

Based on the results of the ultimate analysis, considering the coal blending ratios studied, the 25 % w/w Semirara coal blended with 75 % w/w torrefied sugarcane (*S. officinarum*) bagasse is considered to be the most favorable.

3.3 Calorific Value

The calorific value of a fuel is the number of heat units evolved when unit mass (or unit volume in the case of a gas) of a fuel is completely burned and the combustion products are cooled to 298 K. This definition of calorific value includes the provision that the products of combustion are cooled to 298 K which means the sensible heat and the latent heat of condensation of the water produced during combustion are included in the heat liberated. Therefore, the calorific value of the fuel is designated as 'gross calorific value (GCV)' or 'high heating values (HHV)' [28].

The energy analysis of torrefied sugarcane (*Saccharum officinarum*) bagasse, Semirara coal, and its blends (% w/w) – 25, 50, and 75 were studied by determining the calorific value experimentally and calculating the calorific value theoretically using existing correlations found in literature applicable for both coal and biomass based on proximate and ultimate analyses as summarized in Tables 6 and 7, respectively. Also presented in these tables are the average absolute deviations (AAD) to describe variation in a data set for experimental and calculated values, and the coefficient of determination, (R^2) to measure the goodness of fit of the experimental values to the model. Here, as shown in Equation (13), the AAD is evaluated as

$$\text{AAD} = \frac{1}{n} \times \sum_{i=1}^{n} \left| \left(\varepsilon_{\text{cald}} - \varepsilon_{\text{expt}} \right) \right|_i , \qquad (13)$$

where n is the number of data points and ($\varepsilon_{\text{cald}}$ and $\varepsilon_{\text{expt}}$) are calculated and experimental values, respectively.

The carbon content in ultimate analysis was a predictive basis of the calorific value. And evidently, calorific values and the amount of carbon content show a direct proportionality since the calorific value of blended torrefied sugarcane (*S. officinarum*) bagasse and Semirara coal increases with increasing biomass concentration from 16.17 MJ/kg to 23.20 MJ/kg or by 43.48%.

Table 6. Summary results of the calorific values (experimental) for torrefied sugarcane (*Saccharum officinarum*) bagasse, Semirara coal, and its blends (% w/w) – 25, 50, and 75 and calorific values (calculated) from literature correlations based on proximate analysis.

Concentration (% w/w) of Biomass	$CV_{exp.}$, $MJ \cdot kg^{-1}$	Literature Correlations $CV_{cald.}$, $MJ \cdot kg^{-1}$				
		[14]	[15]	[16]	[17]	[18]
0	16.17	14.91	12.95	14.83	13.24	12.95
25	18.57	13.91	12.07	15.15	13.83	12.07
50	20.16	13.21	11.44	15.43	14.25	11.44
75	21.82	11.91	10.31	15.05	13.94	10.31
100	23.20	11.18	9.64	15.64	14.64	9.64
OAAD (%)	-	32.92	41.88	22.77	29.18	41.88
R^2	-	0.983	0.984	0.587	0.803	0.984

Table 7. Summary results of the calorific values (experimental) for torrefied sugarcane (*Saccharum officinarum*) bagasse, Semirara coal, and its blends (% w/w) – 25, 50, and 75 and calorific values (calculated) from literature correlations based on ultimate analysis.

Concentration (% w/w) of Biomass	CV_{exp}, $MJ \cdot kg^{-1}$	Literature Correlations CV_{calc}, $MJ \cdot kg^{-1}$					
		[19]	[20]	[21]	[22][b]	[22][c]	[23]
0	16.17	14.03	17.51	16.62	16.57	15.09	15.92
25	18.57	15.83	19.12	17.68	18.31	16.21	18.26
50	20.16	18.19	21.05	19.06	20.41	17.72	21.07
75	21.82	20.06	22.63	20.14	22.06	18.87	23.29
100	23.20	21.55	23.96	21.03	23.50	19.84	25.22
AAD (%)	-	10.59	4.53	6.02	1.49	11.90	4.64
R^2	-	0.989	0.991	0.989	0.992	0.988	0.992

Considering the calorific values calculated from literature correlations based on for both proximate and ultimate analyses, the model presented by Sheng and Azevedo [22][b] has the lowest AAD value of 1.49% among the eleven correlations, at the same time the experimental values best fit with that same model, as shown in Figure 3, having an R^2 value of 0.992. The calorific value is the key parameter to evaluate the fuel quality of a special biomass material in energetic application [29]. Higher calorific value is desirable to achieve higher rank of coal. Blended biomass-coal can be predicted to be able to keep pace with pure coal in terms of calorific value.

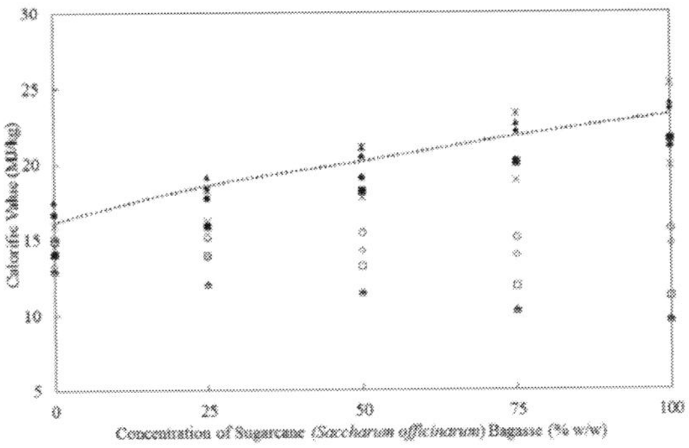

Fig. 3. Plot of experimental (dotted line) calorific values of blended Semirara coal and torrefied sugarcane (*Saccharum officinarum*) bagasse fitted to various correlations applicable for both coal and biomass based on: proximate analysis – Kavšek et al. (□), Kieseler et al. (△), Majumder et al. (○), Mesroghli et al. (◇), Parikh et al. (—); and ultimate analysis – Channiwala & Parihk (■), Friedl et al. (▲), Jenkins et al. (●), Sheng & Azevedo[b] (◆), Sheng & Azevedo[c] (×), Tillman (✱).

4 Conclusion

This paper aimed to establish recommended blending ratios for Semirara coal, a sub-bituminous type of coal from Semirara Island in the province of Antique, Philippines and torrified sugarcane (*Saccharum officinarum*) bagasse thru coal blending technique in order to improve the quality of indigenous coal reserves while concurrently reducing and reusing agricultural wastes. The torrefied sugarcane (*S. officinarum*) bagasse, Semirara coal, and its blends (% w/w) – 25, 50, and 75 were subjected to proximate and ultimate analyses tests, and calorific value determination to characterize and understand the physical conditions and coal properties during combustion.

Using the linear additive rule, it can be established that Semirara coal blended with sugarcane (*S. officinarum*) bagasse were additive based on its proximate and ultimate analyses parameters.

Blending torrefied sugarcane (*S. officinarum*) with Semirara coal would reduce the moisture content and fixed carbon by 77.84% and 63.91%, respectively, while increasing the volatile matter by 74.54% and the ash content by 25.56%. On the other hand, the carbon content increased by 52.86% which is a good indicative property of coal ranks; the hydrogen and sulfur content decreased by 45.12% and 59.74%, respectively which imply that the coal blend improved to higher rank and became environment-friendly. Remarkably, the calorific value was enhanced by 43.48%.

In terms of the coal blending ratios used in this study, the 75 % w/w Semirara coal blended with 25 % w/w torrefied sugarcane (*S. officinarum*) bagasse is considered to be the most favorable based on the results of the proximate analysis. Whereas for the ultimate analysis, it is the 25 % w/w Semirara coal blended with 75 % w/w torrefied sugarcane (*S. officinarum*) bagasse. Among the eleven correlations considered, the model presented by Sheng and Azevedo [22][b] best fit with the experimental values with R^2 value of 0.992 and AAD value of 1.49%.

Therefore, blending torrefied sugarcane (*S. officinarum*) bagasse with Semirara coal would generally improve its quality in terms of its combustion properties thereby making these combinations of coal and biomass advantageous.

This research was supported by the Department of Science and Technology – Engineering Research and Development for Technology (DOST – ERDT) thru a scholarship/research grant coordinated by the Mapúa University – School of Graduate Studies (GS).

References

1. M. S. Roni, S. Chowdhury, S. Mamun, M. Marufuzzaman, W. Lein, S. Johnson, Renew. Sust. Energ. Rev. **78**, 1089-1101 (2017)
2. Y. Zhai, L. Mo, M. Rawlins, *ADB Sustainable Development Working Paper Series: The Impact of Nationally Determined Contributions on the Energy Sector: Implications for ADB and Its Developing Member Countries*, **54** (2018)
3. J. R. Sadullo, *Annual Report: Semirara Mining and Power Corporation – SEC Form 17-A* (2016)
4. S. P. Schweinfurth, R. B. Finkelman, *Coal – A Complex Natural Resource: An overview of factors affecting coal quality and use in the United States*, **1143** (2002)
5. B. Bartok, A. F. Sarofina, *Fossil Fuel Combustion: A Source Book* (John Wiley & Sons, Inc., 1991)
6. Z. Luo, M. Agraniotis, *Low-rank Coals for Power Generation, Fuel and Chemical Production* (Woodhead Publishing, 2017)
7. S. Lee, S. Kim, D. Chun, H. Choi, J. Yoo, *Upgrading and advanced cleaning technologies for low-rank coals – Low-rank Coals for Power Generation, Fuel and Chemical Production* (Woodhead Publishing, 2017)
8. European Commission – Infrastructure for Spatial Information in Europe (INSPIRE). *Fossil Fuel: Low Rank Coal.* Retrieved from http://inspire.ec.europa.eu/codelist/FossilFuelValue/lowRankCoal
9. A. S. F. Chui, J. Z. Sengson, *Philippine Biomass Utilization: A Country Paper Report* (2004)
10. T. A. Mamvura, G. Pahla, E. Muzenda, SAJCE **25**, 1-12 (2018)
11. C. M. S. da Silva, A. D. C. O. Carneiro, B. R. Vital, C. G. Figueiró, L. de Freitas Fialho, M. A. de Magalhães, A. G. Carvalho, W. L. Cândido, Renew. Sust. Energ. Rev. **82**, 2426-2432 (2018)
12. ASTM D3172-13, Standard Practice for Proximate Analysis of Coal and Coke, ASTM International, West Conshohocken, PA, 2013, www.astm.org
13. Y. D. Singh, P. Mahanta, U. Bora, Renew. Energ. **103**, 490-500 (2017)

14. D. Kavšek, A. Bednárová, M. Biro, R. Kranvogl, D. B. Vončina, E. Beinrohr, CEJC **11**, 1481-1491 (2013)
15. S. Kieseler, Y. Neubauer, N. Zobel, Energ. Fuel. **27**, 908-918 (2013)
16. A. K. Majumder, R. Jain, P. Banerjee, J. P. Barnwal, Fuel **87**, 3077-3081 (2008)
17. S. Mesroghli, E. Jorjani, S.C. Chelgani, Int. J. Coal Geol. **79**, 49-54 (2009)
18. J. Parikh, S. A. Channiwala, G. K. Ghosal, Fuel **84**, 487-494 (2005)
19. S. A. Channiwala, P. P. Parikh, Fuel **81**, 1051-1063 (2002)
20. A. Friedl, E. Padouvas, H. Rotter, K. Varmuza Anal. Chim. Acta **544**, 191-198 (2005)
21. B. Jenkins, L. L. Baxter, T. R. Miles Jr, T. R. Miles, Fuel Process. Technol. **54**, 17-46 (1998)
22. C. Sheng, J. L. T. Azevedo, Biomass Bioenergy **28**, 499-507 (2005)
23. D. A. Tillman, *Wood as an Energy Resource* (Elsevier 2012)
24. S. Li, N. Whitely, W. Xu, W. P. Pan, "*Characterization of Coal by Thermal Analysis Methods*" (2005)
25. M. A. Sukiran, F. Abnisa, W. M. A. W. Daud, N. A. Bakar, S. K. Loh, Energy Convers. Manag. **149**, 101-120 (2017)
26. C. Ö. Karacan, R. A. Olea, J. Geochem. Explor. **186**, 36-49 (2018)
27. M. A. Rasheed, P. S. Rao, A. Boruah, S. Z. Hasan, A. Patel, V. Velani, K. Patel, Geosci. J. **5**, 113-119 (2015)
28. H. Liu, *Biomass Fuels for Small and Micro Combined Heat and Power (CHP) Systems: Resources, Conversion and Applications. In Small and Micro Combined Heat and Power (CHP) Systems* (Woodhead Publishing 2011)
29. A. Özyuğuran, S. Yaman, Energy Proced **107**, 130-136 (2017)

Physical and Chemical Properties of Philippine Coal Blended with Torrefied Biomass from Rice (*Oryza sativa*) Straw

Rose Ann P. Lomeda-De Mesa[1,4], *Allan N.* Soriano[3], *Ariziel Ruth D.* Marquez[1,2], and *Adonis P.* Adornado[1,2,4,*]

[1]School of Graduate Studies, Mapúa University, Manila 1002, Philippines
[2]School of Chemical, Biological, and Materials Engineering and Sciences, Mapúa University, Manila 1002, Philippines
[3]Department of Chemical Engineering, Gokongwei College of Engineering, De La Salle University, 2401 Taft Avenue, Manila 1004, Philippines
[4]General Education Department, Colegio de Muntinlupa, Mayor J. Posadas Avenue, Sucat 1770, Muntinlupa City, Metro Manila, Philippines

Abstract. Looking at the feasibility of using rice (*Oryza sativa*) straw locally known as "*dayami*", which the Philippines is quantitatively abundant, for power generation could help farmers reduce production cost and minimize the postharvest waste. Hence, this study's objective is to provide a detailed knowledge of the physical and chemical properties of Philippine coal blended with torrefied rice (*O. sativa*) straw to efficiently use biomass for energy production. Proximate, ultimate, and calorific value analyses showed that rice (*O. sativa*) straw like other biomass resources are appropriate to meet the requirement of thermochemical process. When compared to pure coal, agricultural residues biomass like rice (*O. sativa*) straw has lower moisture and ash content, likewise nitrogen and sulfur are very low as well. Generally, blending Philippine coal with torrefied rice (*O. sativa*) straw would improve its quality in terms of its combustion properties thereby making these combinations of coal and biomass advantageous.

1 Introduction

According to the Philippine Rice Research Institute (PhilRice), the country produces 15.2 million tons of rice (*Oryza sativa*) that leave behind 11.3 million tons of rice (*O. sativa*) straw a year which farmers should not burn in the open field after harvest for environmental and economic reasons [1]. For that reason, there is a need to adopt proper approaches to reduce and reuse agricultural waste. In the Philippines, one coal resources is in Semirara Island located in Caluya, Antique, with an estimated total coal reserve of around 170 million metric tons [2]. However, Philippine coals are black in color, dull (not shiny), have higher moisture and volatile matter, and lower sulfur and fixed carbon content compared to other types of coal, referred to as sub-bituminous [3,4]. Moreover, it is classified as second lowest rank of coals due to their lower calorific value [5-7].

Utilizing and improving the quality of Philippine indigenous coal is getting more important in order to sustain the country's high dependence on coal and to ensure that the growing energy demand can also be addressed. In addition to that, the prices of oil and gas are kept on increasing, thus it is important to develop a new energy system that will serve as an alternative to the non-renewable, highly centralized, and undiversified system by utilizing local sources of energy such as rice (*O. sativa*) straw used as feedstock or raw material in generating electricity [8,9].

Thus, the main purpose of this work is to provide a detailed knowledge of the physical and chemical properties of Philippine coal blended with torrefied rice (*O. sativa*) straw through proximate, ultimate, and calorific value analyses to efficiently use biomass for energy production. This sort of study could serve to create a database of biomass fuels which would support decision making in terms of energy conversion technology selection and operating condition setting.

2 Methodology

2.1. Preparation of Rice (*Oryza sativa*) Straw

For this study, 3 kilograms of raw rice (*Oryza sativa*) straw were collected from a rice field in the Town of Calaca in the province of Batangas, Philippines. Representative rice (*O. sativa*) samples were taken to the Bureau of Plant Industry (BPI) for plant identification, species authentication, and certification. The raw rice (*O. sativa*) straw was reduced in size in the range of 1.0 to 1.5 cm, and then dried using a Binder Drying Oven ED 53 at 105°C for 24 hours prior torrefaction. The weights of the raw and the oven-dried rice (*O. sativa*) straw were recorded, and then the percentage of moisture removed from drying was calculated.

* Corresponding author: apadornado@mapua.edu.ph

2.2 Torrefaction of Rice (*Oryza sativa*) Straw

Biomass torrefaction is a mild thermochemical process that generally performed in inert atmosphere in the temperature range of 200 to 300°C for several minutes or hours. This treatment gained interest as it promotes an increase in the energy content of biomass to levels equal to and sometimes above that of coal [10], making it a more attractive and competitive source within the primary energy matrix [11].

The oven-dried rice (*Oryza sativa*) straw was torrefied using an electric Ney Vulcan 3-1750 Box Muffle Furnace equipped with thermocouple, slide out heating elements, combined fiber, and firebrick insulation for quick cooling and three-stage programmable system of temperature regulation. The 1,750 cubic inches inner volume capacity of the furnace was connected to the outer atmosphere through a tube that provides access of oxygen into the torrefaction zone and discharge of gaseous pyrolysis products.

For each run, approximately 200 grams of oven-dried rice (*O. sativa*) straw were placed into unsealed 110.45 cubic inches ceramic vessels to provide free access of oxygen. The oven-dried rice (*O. sativa*) straw was torrified at 300°C (heating rate of 5°C/min) for 30 minutes under normal atmospheric pressure. The torrefied rice (*O. sativa*) straw was grinded using a Wiley mill and then sifted using a stainless steel laboratory test sieve with mesh screen no. 60 equivalent to an approximate size of 250 µm to homogenize the particle size. Finally, the sifted rice (*O. sativa*) straw were collected in a clean, dry, and sealed bag, and stored in a dessicator until used for blending and characterization. The weights of the oven-dried and the torrified rice (*O. sativa*) straw were recorded, and then the percentage recovery was calculated.

2.3 Proximate Analysis

The proximate analysis performed for the torrefied rice (*Oryza sativa*) straw, Philippine coal, and its blends (% w/w) – 25, 50, and 75 was in accordance to ASTM D3172-13: Standard Practice for Proximate Analysis of Coal and Coke [12]. This practice covers the determination of moisture (MC), volatile matter (VM), and ash (AC) and the calculation of fixed carbon (FC). All measurements performed in this study were carried out in three replicate runs and the mean values were reported.

2.4 Ultimate Analysis

The ultimate analysis for the torrefied rice (*Oryza sativa*) straw, Philippine coal, and its blends (% w/w) – 25, 50, and 75 was performed using LECO CHN628 Series Elemental Determinator and CKIC 5E-S3200 Coulomb Sulfur Analyzer operated using S3200 application program to determine the carbon (C) / hydrogen (H) / nitrogen (N) and the total sulfur (S) content, respectively.

For C/H/N determination, a 0.10 mg of the torrified rice (*O. sativa*) straw was placed in a tin capsule, heated at 980°C with a constant flow of helium enriched with oxygen gas [13].

While for the total S, on average, a 50.45 mg of the torrified rice (*O. sativa*) straw is required, and then covered with a thin layer of tungsten trioxide (WO_3) after weighing in temperature rise period. The furnace was heated to 1,150°C for 30 minutes before analyzing the samples. Also, average values of the triplicate runs were reported for all measurements done in this study.

2.5 Calorific Value (Experimental)

The experimental calorific value (CV) of the torrefied rice (*Oryza sativa*) straw, Philippine coal, and its blends (% w/w) – 25, 50, and 75 was determined using an adiabatic CKIC 5E-AC/PL Calorimeter by completely combusting 1.00 gram of the samples under a pressurized O_2 atmosphere [10]. For all combustion experiments, the samples were charged with oxygen at a pressure of 2.80 to 3.00 MPa and a cylinder pressure of $\geq$ 5 MPa.

2.6 Calorific Value (Calculated)

The values obtained from the proximate and ultimate analyses were used to calculate the calorific value based on existing correlations taken from the literature which are applicable for both coal and biomass based on parameters for proximate and ultimate analyses as shown in Tables 1 and 2, respectively. Thus, it follows that the existing correlations presented are not specific for rice (*Oryza sativa*) straw alone.

Table 1. Correlations for calorific value applicable for both coal and biomass based on proximate analysis.

Study	Correlation for Calorific Value, $MJ \cdot kg^{-1}$	Eq.
Kavšek et al. [14]	CV = 0.4108FC + 0.1934VM - 0.021AC	(1)
Kieseler et al. [15]	CV = 0.3536FC + 0.1559VM - 0.0078AC	(2)
Majumder et al. [16]	CV = -0.03AC - 0.11MC + 0.33VM + 0.35FC	(3)
Mesroghli et al. [17]	CV = 37.777 - 0.647MC - 0.387AC - 0.089VM	(4)
Parikh et al. [18]	CV = 0.3536FC + 0.1559VM - 0.0078AC	(5)

Table 2. Correlations for calorific value applicable for both coal and biomass based on ultimate analysis.

Study	Correlation for Calorific Value, $MJ \cdot kg^{-1}$	Eq.
Channiwala & Parihk [19]	CV = 0.3491C + 1.1783H + 0.1005S - 0.1034O - 0.0151N - 0.0211A	(6)
Friedl et al. [20]	CV = $0.00355C^2$ - 0.232C - 2.23H + 0.0512CH + 0.131N + 20.6	(7)
Jenkins et al. [21]	CV = -0.763 + 0.301C + 0.525H + 0.064O	(8)
Sheng & Azevedo et al. [22][b]	CV = 0.3259C + 3.4597	(9)
Sheng & Azevedo et al. [22][c]	CV = -1.3675 + 0.3137C + 0.7009H + $0.0318O^{*a}$	(10)
Tillman [23]	CV = 0.4373C -1.6701	(11)

[a] O^* is the sum of the contents of oxygen and other elements in the organic matter (O^* = 100-C-H-Ash).

3 Results and Discussion

The results for the proximate analysis, ultimate analysis, and calorific value for the torrefied rice (*Oryza sativa*) straw, Philippine coal, and its blends (% w/w) – 25, 50, and 75 were summarized in Tables 3, 4, and 5, respectively.

Table 3. Results for the proximate analysis – moisture (MC), ash (AC), volatile matter (VM), and fixed carbon (FC) on the basis of concentration (% w/w) of rice (*Oryza sativa*) straw.

Concentration (% w/w) of Biomass	MC (%)	AC (%)	VM (%)	FC (%)
0	13.04	36.03	24.19	26.74
25	9.87	34.42	25.60	30.11
50	6.54	32.76	27.19	33.51
75	4.62	31.69	27.64	36.05
100	2.00	30.53	29.03	38.44

Table 4. Results for the ultimate analysis – carbon (C), hydrogen (H), nitrogen (N), and sulfur (S) on the basis of concentration (% w/w) of rice (*Oryza sativa*) straw.

Concentration (% w/w) of Biomass	C (%)	H (%)	N (%)	S (%)
0	40.22	4.61	9.80	0.77
25	40.86	4.31	11.22	0.71
50	40.64	3.94	11.01	0.55
75	38.21	2.95	10.39	0.40
100	36.80	2.53	10.49	0.35

Table 5. Results for calorific value and energy yield on the basis of concentration (% w/w) of rice (*Oryza sativa*) straw.

Concentration (% w/w) of Biomass	Calorific Value ($MJ \cdot kg^{-1}$)	Energy Yield (%)
0	16.17	-
25	16.37	60.80
50	15.61	58.00
75	15.25	56.65
100	14.57	54.13

3.1 Proximate Analysis

Generally, any two coals cannot just be blended. To decide to blend or not, it is very important to understand the composition of the coals that are to be blended. The linear additive rule shown in Equation (12) is used to estimate the theoretical composite value of the coal blends to show the relationship of the coal properties and the amount of that coal in the blend, where M is the composite value of the parameter being estimated and x is the weight fraction of the components in the blend.

$$M = x_b M_b + (1 - x_b) M_a \quad (12)$$

A property can be considered additive when the physical property of the blend can be predicted by the relative amounts of component coals and their physical properties. The applicability of the additive property for coal blending was determined using the coefficient of determination, R^2 value of at least 0.9000 [24]. Based on Figure 1, comparing the experimental values from the calculated values for moisture (MC), ash (AC), volatile matter (VM), and fixed carbon (FC) based on the linear additive rule, it showed acceptable results with minimal deviations with overall R^2 value of 0.9618. Thus, it can be established that Philippine coal blended with rice (*Oryza sativa*) straw were additive based on its proximate analysis parameters.

The proximate analysis involves the quantitative determination of moisture content (MC), ash content (AC), volatile matter (VM), and fixed carbon (FC). The results for the proximate analysis for the torrefied rice (*O. sativa*) straw, Philippine coal, and its blends (% w/w) – 25, 50, and 75 was shown in Table 3. The trends of these values were also illustrated in Figure 1 on the basis of increasing concentration (% w/w) of rice (*O. sativa*) straw.

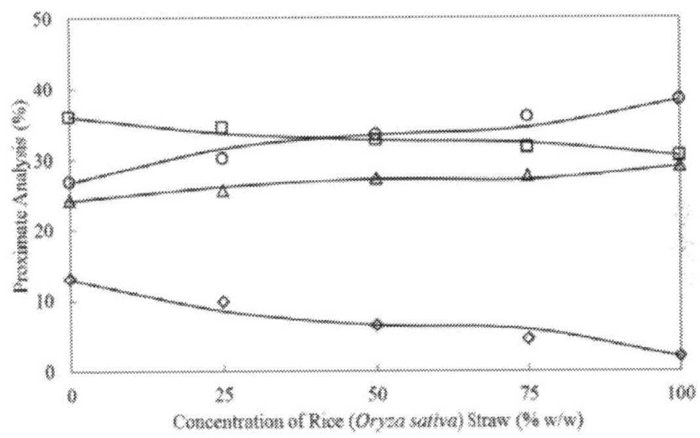

Fig. 1. Comparison of experimental values for proximate analysis parameters – moisture (◇), ash (□), volatile matter (△), and fixed carbon (○) versus calculated values (solid lines) based on linear additive rule.

It was observed that an increase in the concentration of torrefied rice (*O. sativa*) straw in the coal blend caused a decrease in the moisture content (MC) by as much as 84.66%. An increase in the moisture content of coal results in an increase in heat loss due to evaporation and superheating of vapor. Therefore, an increase in the concentration of torrefied rice (*O. sativa*) straw in the coal blend is a good indication that it averts the presence of fungi and decreases other biological activities upon storing. High moisture content can affect the overall energy conversion upon combustion because it influences the calorific value of the given fuel. The higher moisture content, the less energy per kilogram is released that causes lower calorific value [25]. This claim could be supported by the relationship of moisture content and calorific value of other biomass resources studied by the researchers such as for sugarcane (*Saccharum officinarum*) bagasse and coconut (*Cocos nucifera*) husk. However, for rice (*O. sativa*) straw, a decrease in the moisture content in the coal blend still corresponds to a decrease in the calorific value as shown in Table 5 due to the decrease in the carbon content based on ultimate analysis, which is also a predictive basis for calorific value.

Ash (AC) is the inorganic residue after combustion. One characteristic of a good fuel is having little amount of ash during combustion. Small amount of ash residues signifies the maximum utilization of the fuel used. High amount of ash affects combustion efficiency and boiler efficiency because it causes clinkering and slagging. Looking at Table 3, it can be observed that an increase in the amount of torrefied rice (*O. sativa*) straw in the coal blend would decrease the amount of ash by 15.27%.

High volatile matter (VM) indicates that the fuel can be easily ignited and subsequently oxidized. High volatile matter also contributes to better efficiency for burning during combustion. It influences the secondary air requirement, distribution aspects, and secondary oil supports. An increase in the amount of torrefied rice (*O. sativa*) straw in the coal blend corresponds to an increase in volatile matter by 20.01%.

Fixed carbon (FC) is the solid combustible residue that remains after a biomass particle is heated and the volatile matter is expelled. The amount of fixed carbon and volatile combustible matter directly contribute to the heating value of coal. Fixed carbon acts as a main heat generator during burning. High fixed carbon in biomass increases the char formation [25]. An increase in the amount of torrefied rice (*O. sativa*) straw in the coal blend also increases the amount of fixed carbon by 43.75%.

Lastly, in terms of the coal blending ratios used in this study, the 25 % w/w Philippine coal blended with 75 % w/w torrefied rice (*O. sativa*) straw is considered to be the most favorable based on the results of the proximate analysis.

3.2 Ultimate Analysis

Shown in Figure 2 is the comparison of experimental values versus the calculated values based on the linear additive rule for ultimate analysis – carbon (C), hydrogen (H), nitrogen (N), and sulfur (S). Just like in proximate analysis, the values are within the allowable results with slight deviations with overall R^2 value of 0.9637. Thus, it follows that Philippine coal blended with rice (*Oryza sativa*) straw were also additive based on its ultimate analysis parameters.

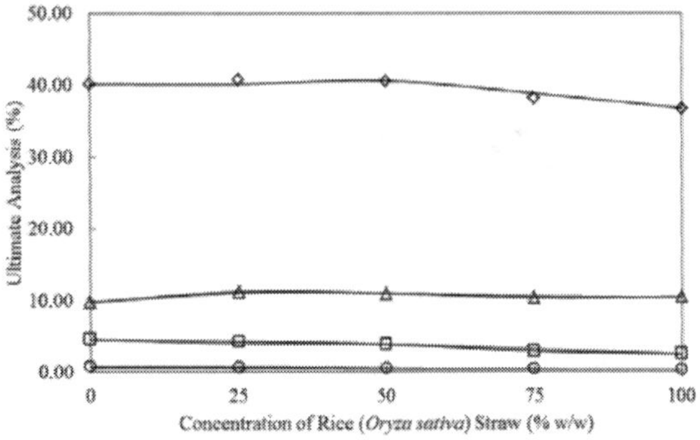

Fig. 2. Comparison of experimental values for ultimate analysis parameters – carbon (◇), hydrogen (□), nitrogen (△), and sulfur (○) versus calculated values (solid lines) based on linear additive rule.

The ultimate analysis determines the elemental composition of the coal including moisture, ash, carbon (C), hydrogen (H), nitrogen (N), sulfur (S), and oxygen (O) (by difference). It is valuable in determining the quality of air required for combustion and composition of combustion gases. The results for the ultimate analysis for the torrefied rice (*O. sativa*) straw, Philippine coal, and its blends (% w/w) – 25, 50, and 75 was shown in Table 4. The trends of these values were also illustrated in Figure 2 on the basis of increasing concentration (% w/w) of rice (*O. sativa*) straw.

Carbon content is a good indicative property of coal ranks. The highest rank coals have the highest carbon contents and the lowest oxygen contents. Ratios of these elements can indicate the rank of the coal and its coalification degree [26]. It is also important to measure the sulfur content in coal samples to evaluate the potential sulfur emissions from coal combustion. Increasing the amount of torrefied rice (*O. sativa*) straw in the coal blend decreases the carbon content by 8.50%.

Meanwhile, the hydrogen content for all coal blends decreases by 45.12% with increasing amount of torrefied rice (*O. sativa*) straw. The increased content of hydrogen is normally more of a characteristic of low rank coals, while the decreased values are commonly more typical for higher-rank coals [27]. Based on this, it could be said that a decrease in hydrogen content indicates that the quality of Philippine coal blended torrefied rice (*O. sativa*) straw improved to higher rank.

Nitrogen content for all coal blends is approximately the same with 10.58% on average. Nitrogen is an inert and incombustible gas and it does not contribute any useful property to the combustion of coal. It is generally found in the organic fraction of the coal wherein upon combustion, it is emitted as nitrogen oxide (NO_x) in the flue gas.

Sulfur content for all coal blends decreases by 54.55% with increasing amount of torrefied rice (*O. sativa*) straw. Presence of sulfur in coal is harmful for use in metallurgy as it transfers to the metal and adversely affects the property of metal. Also, oxidation products of sulfur such as sulfur dioxide (SO_2) and sulfur trioxide (SO_3) have corrosive effect on the equipment and causes atmospheric pollution. Hence the presence of sulfur in coal is undesirable. A decrease in the sulfur content of Philippine coal upon coal blending with torrefied rice (*O. sativa*) straw is an indication that an environment-friendly approach have been taken.

Based on the results of the ultimate analysis, considering the coal blending ratios studied, the 25 % w/w Philippine coal blended with 75 % w/w torrefied rice (*O. sativa*) straw is considered to be the most favorable.

3.3 Calorific Value

The calorific value of a fuel is the number of heat units evolved when unit mass (or unit volume in the case of a gas) of a fuel is completely burned and the combustion products are cooled to 298 K. This definition of calorific value includes the provision that the products of combustion are cooled to 298 K which means the sensible heat and the latent heat of condensation of the water produced during combustion are included in the heat liberated. Therefore, the calorific value of the fuel is designated as 'gross calorific value (GCV)' or 'high heating values (HHV)' [28].

The energy analysis of torrefied rice (*Oryza sativa*) straw, Philippine coal, and its blends (% w/w) – 25, 50, and 75 were studied by determining the calorific value experimentally and calculating the calorific value theoretically using existing correlations found in literature applicable for both coal and biomass based on proximate and ultimate analyses as summarized in Tables 6 and 7, respectively. Also presented in these tables are the average absolute deviations (AAD) to describe variation in a data set for experimental and calculated values, and the coefficient of determination, (R^2) to measure the goodness of fit of the experimental values to the model. Here, as shown in Equation (13), the AAD is evaluated as

$$\text{AAD} = \frac{1}{n}\times\sum_{i=1}^{n}\left|\left(\varepsilon_{\text{cald}}-\varepsilon_{\text{expt}}\right)\right|_i , \quad (13)$$

where n is the number of data points and ($\varepsilon_{\text{cald}}$ and $\varepsilon_{\text{expt}}$) are calculated and experimental values, respectively.

The carbon content in ultimate analysis was a predictive basis of the calorific value. And evidently, calorific values and the amount of carbon content show a direct proportionality since the calorific value of blended torrefied rice (*O. sativa*) straw and Philippine coal decreases with increasing biomass concentration (decreasing carbon content) from 16.17 to 14.57 MJ/kg or by 9.89%.

Table 6. Summary results of the calorific values (experimental) for torrefied rice (*Oryza sativa*) straw, Philippine coal, and its blends (% w/w) – 25, 50, and 75 and calorific values (calculated) from literature correlations based on proximate analysis.

Concentration (% w/w) of Biomass	$CV_{exp.}$, MJ·kg^{-1}	Literature Correlations $CV_{cald.}$, MJ·kg^{-1}				
		[14]	[15]	[16]	[17]	[18]
0	16.17	14.91	12.95	14.83	13.24	12.95
25	16.37	16.60	14.37	16.87	15.79	14.37
50	15.61	18.34	15.83	19.00	18.45	15.83
75	15.25	19.49	16.81	20.28	20.06	16.81
100	14.57	20.76	17.88	21.90	22.08	17.88
OAAD (%)	-	6.96	8.70	4.77	6.00	8.70
R^2	-	0.992	0.993	0.579	0.782	0.993

Table 7. Summary results of the calorific values (experimental) for torrefied rice (*Oryza sativa*) straw, Philippine coal, and its blends (% w/w) – 25, 50, and 75 and calorific values (calculated) from literature correlations based on ultimate analysis.

Concentration (% w/w) of Biomass	CV_{exp}, MJ·kg^{-1}	Literature Correlations CV_{calc}, MJ·kg^{-1}					
		[19]	[20]	[21]	[22][b]	[22][c]	[23]
0	16.17	14.03	17.51	16.62	16.57	15.09	15.92
25	16.37	14.08	17.92	16.54	16.78	15.12	16.20
50	15.61	13.49	17.89	16.35	16.70	14.86	16.10
75	15.25	11.06	17.47	15.36	15.91	13.55	15.04
100	14.57	9.91	17.37	14.83	15.45	12.91	14.42
AAD (%)	-	2.05	0.87	0.29	1.26	2.44	0.99
R^2	-	0.995	0.996	0.996	0.995	0.995	0.996

Considering the calorific values calculated from literature correlations based on for both proximate and ultimate analyses, the model presented by Jenkins et al. [21] has the lowest AAD value of 0.29% among the eleven correlations, at the same time the experimental values best fit with that same model, as shown in Figure 3, having an R^2 value of 0.996. The calorific value is the key parameter to evaluate the fuel quality of a special biomass material in energetic application [29]. Higher calorific value is desirable to achieve higher rank of coal. Blended biomass-coal can be predicted to be able to keep pace with pure coal in terms of calorific value.

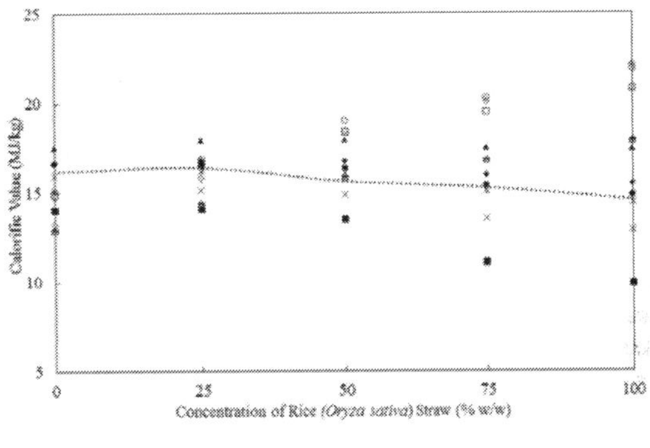

Fig. 3. Plot of experimental (dotted line) calorific values of blended Philippine coal and torrefied rice (*Oryza sativa*) straw fitted to various correlations applicable for both coal and biomass based on: proximate analysis – Kavšek et al. (□), Kieseler et al. (△), Majumder et al. (○), Mesroghli et al. (◇), Parikh et al. (—); and ultimate analysis – Channiwala & Parihk (■), Friedl et al. (▲), Jenkins et al. (●), Sheng & Azevedo[b] (◆), Sheng & Azevedo[c] (×), Tillman (✱).

4 Conclusion

This paper aimed to provide a detailed knowledge of the physical and chemical properties of Philippine coal blended with torrefied rice (*Oryza sativa*) straw through proximate, ultimate, and calorific value analyses.

Using the linear additive rule, it can be established that Philippine coal blended with torrefied rice (*O. sativa*) straw were additive based on its proximate and ultimate analyses parameters.

Blending Philippine coal with torrefied rice (*O. sativa*) straw would reduce the moisture and ash content by 84.66% and 15.27%, respectively, while increasing the volatile matter by 20.01% and the fixed carbon content by 43.75%. On the other hand, the carbon content decreased by 8.50%, lowering the calorific value by 9.89%.; the hydrogen and sulfur content decreased by 45.12% and 54.55%, respectively which imply that the coal blend improved to higher rank and became environment-friendly.

In terms of the coal blending ratios used in this study, the 25 % w/w Philippine coal blended with 75 % w/w torrefied rice (*O. sativa*) straw is considered to be the most favorable based on the results for both proximate and ultimate analysis. Among the eleven correlations considered, the model presented by Jenkins et al. [21]

best fit with the experimental values with R^2 value of 0.996 and AAD value of 0.29%.

Therefore, blending Philippine coal with torrefied rice (*O. sativa*) straw would generally improve its quality in terms of its combustion properties thereby making these combinations of coal and biomass advantageous.

This research was supported by the Department of Science and Technology – Engineering Research and Development for Technology (DOST – ERDT) thru a scholarship/research grant coordinated by the Mapúa University – School of Graduate Studies (GS).

References

1. Philippine Rice Research Institute (PhilRice). *PhilRice to study rice straw as power source.* Retrieved from http://www.philrice.gov.ph/philrice-to-study-rice-straw-as-power-source/
2. J. R. Sadullo, *Annual Report: Semirara Mining and Power Corporation – SEC Form 17-A* (2016)
3. S. P. Schweinfurth, R. B. Finkelman, *Coal – A Complex Natural Resource: An overview of factors affecting coal quality and use in the United States*, **1143** (2002)
4. B. Bartok, A. F. Sarofina, *Fossil Fuel Combustion: A Source Book* (John Wiley & Sons, Inc., 1991)
5. Z. Luo, M. Agraniotis, *Low-rank Coals for Power Generation, Fuel and Chemical Production* (Woodhead Publishing, 2017)
6. S. Lee, S. Kim, D. Chun, H. Choi, J. Yoo, *Upgrading and advanced cleaning technologies for low-rank coals – Low-rank Coals for Power Generation, Fuel and Chemical Production* (Woodhead Publishing, 2017)
7. European Commission – Infrastructure for Spatial Information in Europe (INSPIRE). *Fossil Fuel: Low Rank Coal.* Retrieved from http://inspire.ec.europa.eu/codelist/FossilFuelValue/lowRankCoal
8. D. J. Yap (2012). Philippine Daily Inquirer. *PhilRice looks into rice straw as energy source.* Retrieved from https://newsinfo.inquirer.net/259548/philrice-looks-into-rice-straw-as-energy-source
9. Philippine Rice Research Institute (PhilRice). *Don't burn rice straw – PhilRice.* Retrieved from https://www.philrice.gov.ph/dont-burn-rice-straw-philrice/
10. T. A. Mamvura, G. Pahla, E. Muzenda, SAJCE **25**, 1-12 (2018)
11. C. M. S. da Silva, A. D. C. O. Carneiro, B. R. Vital, C. G. Figueiró, L. de Freitas Fialho, M. A. de Magalhães, A. G. Carvalho, W. L. Cândido, Renew. Sust. Energ. Rev. **82**, 2426-2432 (2018)
12. ASTM D3172-13, Standard Practice for Proximate Analysis of Coal and Coke, ASTM International, West Conshohocken, PA, 2013, www.astm.org
13. Y. D. Singh, P. Mahanta, U. Bora, Renew. Energ. **103**, 490-500 (2017)
14. D. Kavšek, A. Bednárová, M. Biro, R. Kranvogl, D. B. Vončina, E. Beinrohr, CEJC **11**, 1481-1491 (2013)
15. S. Kieseler, Y. Neubauer, N. Zobel, Energ. Fuel. **27**, 908-918 (2013)
16. A. K. Majumder, R. Jain, P. Banerjee, J. P. Barnwal, Fuel **87**, 3077-3081 (2008)
17. S. Mesroghli, E. Jorjani, S.C. Chelgani, Int. J. Coal Geol. **79**, 49-54 (2009)
18. J. Parikh, S. A. Channiwala, G. K. Ghosal, Fuel **84**, 487-494 (2005)
19. S. A. Channiwala, P. P. Parikh, Fuel **81**, 1051-1063 (2002)
20. A. Friedl, E. Padouvas, H. Rotter, K. Varmuza Anal. Chim. Acta **544**, 191-198 (2005)
21. B. Jenkins, L. L. Baxter, T. R. Miles Jr, T. R. Miles, Fuel Process. Technol. **54**, 17-46 (1998)
22. C. Sheng, J. L. T. Azevedo, Biomass Bioenergy **28**, 499-507 (2005)
23. D. A. Tillman, *Wood as an Energy Resource* (Elsevier 2012)
24. S. Li, N. Whitely, W. Xu, W. P. Pan, "*Characterization of Coal by Thermal Analysis Methods*" (2005)
25. M. A. Sukiran, F. Abnisa, W. M. A. W. Daud, N. A. Bakar, S. K. Loh, Energy Convers. Manag. **149**, 101-120 (2017)
26. C. Ö. Karacan, R. A. Olea, J. Geochem. Explor. **186**, 36-49 (2018)
27. M. A. Rasheed, P. S. Rao, A. Boruah, S. Z. Hasan, A. Patel, V. Velani, K. Patel, Geosci. J. **5**, 113-119 (2015)
28. H. Liu, *Biomass Fuels for Small and Micro Combined Heat and Power (CHP) Systems: Resources, Conversion and Applications. In Small and Micro Combined Heat and Power (CHP) Systems* (Woodhead Publishing 2011)
29. A. Özyuğuran, S. Yaman, Energy Procedia **107**, 130-136 (2017)

 https://doi.org/10.1051/e3sconf/201912002004

The effect of loading rate to biogas production rate of the 500 liter anaerobic digester operated with continuous system

I Ketut Adi Atmika[1], *Kadek* Sebayuana[2], *Tjokorda Gde Tirta* Nindhia[1,*], *I Wayan* Surata[1], *I Putu Ari* Astawa[3], *Anak Agung Istri Agung Sri* Komaladewi[1]

[1]Study Program of Mechanical engineering, engineering Faculty, Udayana University, Jimbaran, Bali, 80361, Indonesia
[2]Master Student at Study Program of Master in Mechanical Engineering , Engineering Faculty, Udayana University, Jimbaran, Bali, 80361, Indonesia
[3]Study Program of Animal Husbandry, Faculty of Animal Husbandry, Udayana University, Jimbaran, Bali, 80361, Indonesia

Abstract. Conventional anaerobic digester such as fixed dome and floating drum are found having drawback in application in developing country. It was difficult in maintenance and operation. It was also difficult to relocate to the new site of waste processing. The portable anaerobic digester is prepared in this work as a solution. The capacity is about 500 liter so that suitable for home scale organic waste treatment. The material that is used for the digester was 304 stainless steel. The digester is completed wit agitator to optimize the biogas production. A slurry of cow dung (50% cow dung+ 50% water) is use to feed the digester. There are 2 variations of slurry loading rate that were investigated in this work, namely 5 liter slurry/day and 10 liter slurry/day. The biogas production rate is found about 51.7 liter biogas/day if loading with 5 liter slurry/ day. The biogas production rate is found increase significantly to become 82 liter biogas/day if loading with 10 liter slurry/day. The quality of biogas is found better with loading rate 5 liter slurry/day which has average CH4 content about 58.75% vol. comparing the one with loading rate 10liter slurry/day that have average CH4 content about 56.40% vol.

1 Introduction

In many countries biogas technology spread has foundered and/or up to 50% of plants are non-functional. This is linked to inadequate emphasis on maintenance and repair of existing facilities. Many new design of anaerobic digester are introduced with many feature to increase the performance. Attention to establish anaerobic digester in the developing country is increase rapidly [1]. General designs of biogas digester that is used in developing countries for digestion of livestock waste are classified as low-rate digesters, and lacking stirring capability [1].

The disadvantages of low-rate digesters biogas technology such as fixed dome and floating drum are : requires reliable feed source, high construction costs relative to income of users, laborious operation and maintenance, limited lifespan, requires reliable feed source, construction costly[2]. Certain researcher develop small size anaerobic digester but nor for the purpose of portable design [3].

Many of concrete fixed dome type of digester is found difficult to be repair. A static scum-breaking device in the form of nylon net has been incorporated in a fixed-dome biogas plant to control scum formation and to provide a limited amount of stirring [4]. A portable bio-digester should be fabricated with consideration of economic and ergonomic factors with maximum efficiency in production of methane gas [5].

For the case in Vietnam, there no best choice house hold biogas plant model for all the farmers. The existing models have their weaknesses and strengths. More studies are required to find optimal biogas models to the real situation . Such future biogas plant models should take 3 requirements: simple in construction, operation, and maintenance [6].

Many improvements already addressed for the fixed dome and floating drum [7] but portable design was not take in to account in the improvement. Little sized biogas plants solves the problem of the organic waste, reducing its transportation costs and producing green energy [3]. Effort to find best suit biogas digester is still on going, different technologies have been analyzed in order to identify the most suitable for small sized biogas plants [8].

In this work a small size (500 liter) anaerobic digester is introduced and can be operated as continuous system. The design is completed with manual agitator to optimize the result. Stainless steel is used as a material of anaerobic digester for corrosion resistance purpose. The effect of loading rate to the biogas production rate as wellas biogas composition will be investigated

2 Experimental

* Corresponding author: nindhia@yahoo.com

https://doi.org/10.1051/e3sconf/201912002004

The 500 liter portable anaerobic digester was set up at village of Baluk, sub-districts of Negara, districts of Jembrana, In Bali Island, Indonesia. The climate is tropical so that no external heating energy is needed to operate the anaerobic digester. Humidity level is around 85%. The operation temperature was recorded vary between 20-33°C.

The anaerobic digester is made from stainless steel plate (thickness 1.22 mm) to ensure that the digester having high corrosion resistant. The biogas contain hydrogen sulfide (H_2S) [9, 10, 11] that is very corrosive to carbon steel or metals in general.

The anaerobic digester cylinder with diameter 76 cm and length 122 cm was obtained by rolling the stainless steel plate. The anaerobic digester cylinder was mounted horizontally for ergonomic design and easy loading [5]. The manufacture of the stainless steel anaerobic digester utilizes tungsten inert gas (TIG) welding technique.

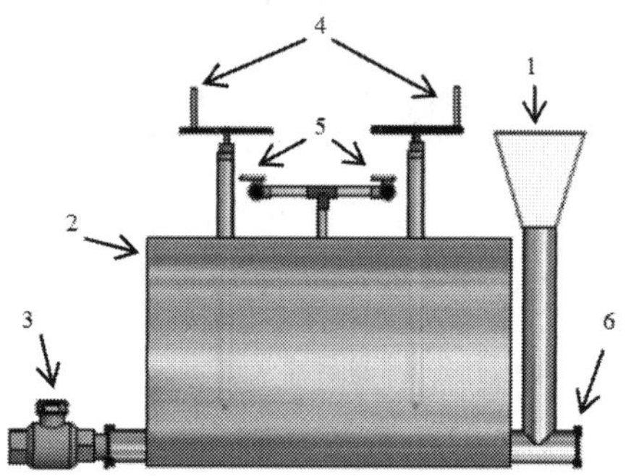

Fig. 1. The design of anaerobic digester. 1. Inlet slurry, 2. Anaerobic tank, 3.Slurry outlet, 4. Manual agitator handle,5. Outflow biogas, and 6. Manhole

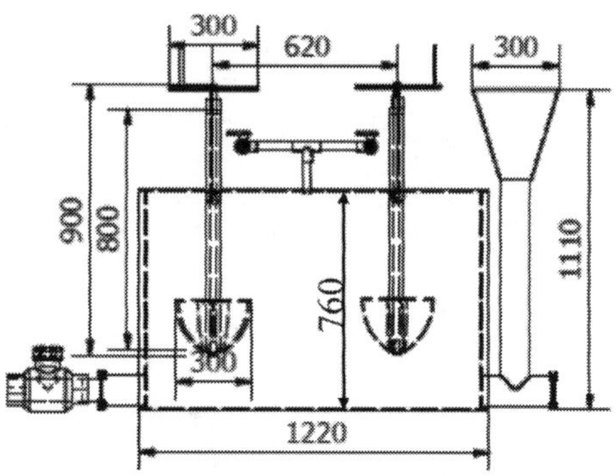

Fig. 2. Section design of 500 liter of anaerobic digester. The unit in millimeter (mm)

The drawing of anaerobic digester is presented in Figure 1, Figure 2, and Figure 3. The main part of the digester are inlet slurry, anaerobic tank, manual agitator handle, outflow biogas, slurry outlet, agitator blade. The manhole is also provided for easy cleaning during maintenance. The total weight of empty digester is about 60 Kg.

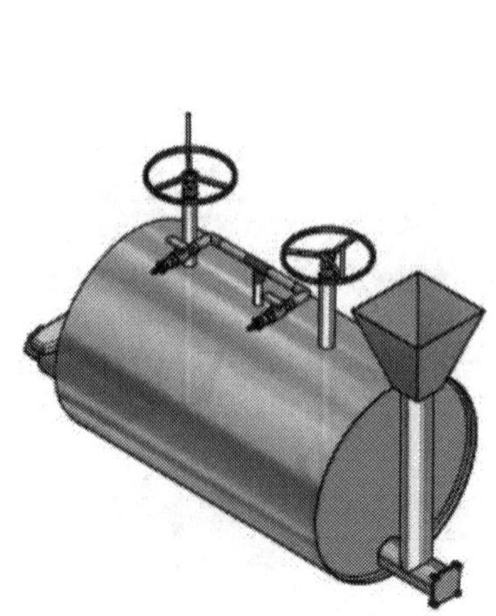

Fig. 3. Three dimension appearance of the anaerobic digester

At the beginning the anaerobic digester is operated as batch system by filling the tank fully with slurry of cow dung (50% cow dung+50% water). The stirring with agitator was conducted for about 10 minutes for 3 times a day (morning, noon, and evening). The anaerobic digestion then is let to produce biogas by batch system until no longer produce biogas again. In this stage the anaerobic digestion process finished byusing batch system.

The research then was continued with the investigation of the performance for continuous system. The first variable was conducted by releasing for about 5 liter slurry from the outlet and about the same amount (5 liter) of new slurry (50% cow dung+50% water) was added from the inlet. The stirring with agitator was conducted for about 10 minute for 3 times a day (morning, noon, and evening). The biogas produced in 30 day then was measured as well as biogas composition. The same process also was conducted for second variable that was by releasing for about 10 liter slurry from the outlet and about 10 liter new slurry (50% cow dung+50% water) was added from the inlet.

3 Results And Discussion

Base on the result from the experiment in this research, it is found that by loading rate 10 liter slurry/day the biogas production rate is higher (around 82.0 L/day) comparing with the one with loading rate 5 liter slurry/day (around 51.7 L/day) as can be seen in Figure 4 as well as Table 1. The biogas production rate from both loading rate are found linear in the range of 30 days (1 month) of biogas production. From this point it is clear that loading rate influent the biogas production rate in the continuous system of 500 liter anaerobic digestion. It is explained in previous publication that indeed loading rate influent the biogas production rate [12].

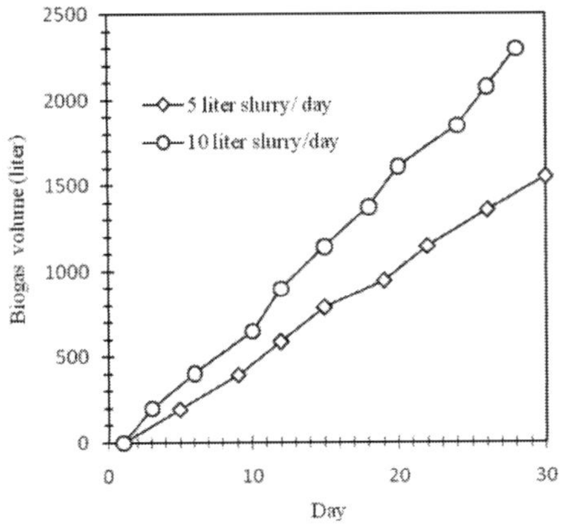

Fig. 4. The effect of loading rate of slurry to biogas production in the 500 liter of continuous anaerobic digestion

Table 1. The effect of loading rate to biogas production rate as well as biogas composition in 30 days (1 month) with continuous system operation.

Loading rate of slurry (liter of slurry/day)	Biogas production rate (liter of biogas/day)	Average biogas composition in 30 days		
		CH_4 (% vol.)	CO_2 (% vol)	H_2S (ppm)
5	51.7	58.75	35.75	42.75
10	82.0	56.40	33.30	106.50

The effect of loading rate to biogas composition can be seen in the Table 1. The result reveal that related with utilization of biogas as a fuel, loading rate 5 liter slurry/day resulting better quality of biogas with CH_4 around 58.75 % vol comparing loading rate 10 liter slurry/day that reach 56.40 % vol. which is a lower value. There are four stage of biological and chemical of anaerobic digestion namely: hydrolysis, acidogenesis, acetogenesis, and methanogenesis as a final stage [12, 13].

Biomass is made up of organic polymers. To access the energy potential of the material , in anaerobic digesters, the bacteria must first broken down these chain in to smaller constituent part such as amino acids, simple sugar, and fatty acids. To break the chain and dissolving the smaller molecules in to the solution is called hydrolysis. Hydrogen and acetate are produced in the hydrolysis and can be used by methanogens directly. volatile fatty acids (VFAs) with a chain length greater than that of acetate must first be catabolised into compounds that can be used by methanogens. The process of acidogenesis results in further breakdown of the remaining components by acidogenic (fermentative) bacteria. Here, VFAs are created, along with carbon dioxide, ammonia, and hydrogen sulfide, as well as other byproducts. Acetogenesis is the third stage of anaerobic digestion. Simple molecules through the acidogenesis phase are further digested by acetogens to produce largely acetic acid, as well as hydrogen and carbon dioxide. The biological process of methanogenesis is the terminal stage of anaerobic digestion. Methanogens use the intermediate products of the preceding stages and convert them into wate r, carbon dioxide and methane that make up the majority of the biogas [12, 13].

CO_2 is produced during stage acidogenesis, acetogenesis, as well as methanogenesis. H_2S is produced during stage of acidogenesis. Methane is produce during stage of methanogenesis [12]. Figure 5 is the result of investigation of CH_4 contents in the biogas in 30 days. It weres found that CH_4 contents in the biogas is about lower for loading rate 10 L/day until reach the 20th day. After that the CH4 contents were found about the same

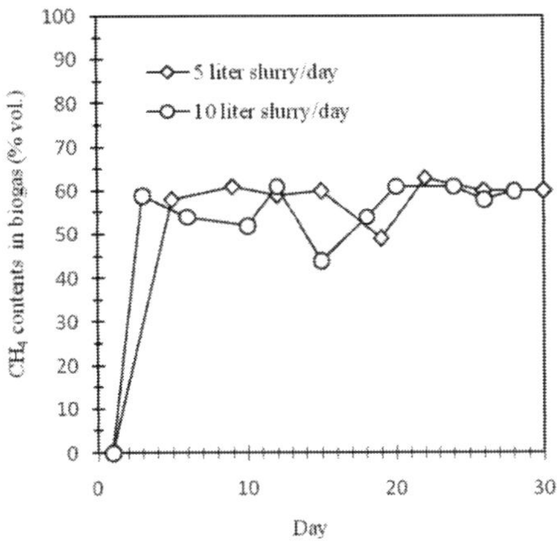

Fig. 5. Methane (CH_4) contents in the biogas in 30 days

The CO_2 contents in the biogas can be seen in Figure 6. Similar to CH4 contents in the biogas, The CO2 contents in the biogas also was found about lower for loading rate 10 L/day comparing with loading rate 5 L/day. After reach the 20th day, the CO_2 contents were found similar.

The H_2S contents in the biogas is presented in Figure 7. The H_2S contents for loading rate 10 L/day was found very much higher comparing loading rate 5 L/day. The H2S contents was found very small for loading rate 5 L/day until reach the 20th day, and after that H2S contents was found similar to loading rate 10 L/day. Seem that for capacity 500 slurry (50% cow dung+50% water) the crucial condition was at the first 20th day. After the day of 2oth the biogas composition were found similar for bot loading rate of slurry

By increasing loading rate to 10 liter slurry/day without adding retention time, this will make the process will stay longer in the 3 first processes (hydrolysis, acidogenesis, acetogenesis) that make the content of CH_4 is lower and H_2S higher (106.50 ppm) than loading rate 5 liter slurry/day. At lower loading rate of 5L/day of slurry also resulting in very low H_2S content (47.50 ppm) which is a good condition. Hydrogen sulfide is produced during stage of acidogenic [12]. Biogas with low content of H_2S in very good as a fuel of the engine

[14] because will reduce the corrosion in the metal component of the engine.

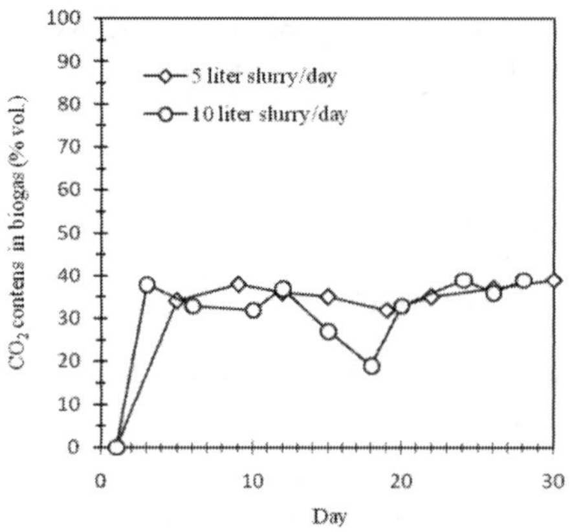

Fig. 6. Carbon dioxide (CO_2) contents in the biogas in 30 days

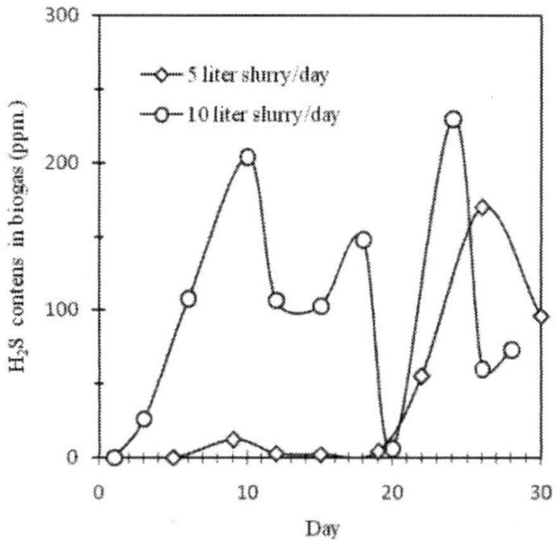

Fig. 7. Hydrogen sulfide (H_2S) contents in the biogas in 30 days

Training and education of householders, is needed in relation to the maintenance of digesters, feedstock suitability and the environmental and potential livelihood benefits of digestate [15]. The result in this work can be informed during training and education of biogas householder.

4 Conclusions

For the 500 liter anaerobic digester made by stainless steel operated in continuous system, the biogas production rate will be around 51.7 L/day if the loading rate of slurry is 5 L/day. The biogas production rate increase to reach 82.0 L/day if the loading rates of the slurry IS increased to 10 L/day. The slurry is made from cow dung (50% cow dung + 50% water).

The quality of biogas is better for loading rate of slurry at 5 L/day with average CH_4 content in 30 days around 58.75%vol. Increasing loading rate of slurry at 10 L/day yield lower average CH_4 content in 30 days to become around 56.40% vol.

The average H_2S content in the biogas is very low (47.50 ppm) for loading rate of slurry 5 L/day and become very much increase at higher loading rate of slurry at 10 L/day which is around 106.50 ppm.

This work is supported from the grant of *Program Pengembangan Desa Mitra* from *Kementrian Riset Teknologi dan Pendidikan tinggi The* Republic of Indonesia with contract number: 493.6/UN14.4.A/PM/2019

References

1. T. Bond, M.R. Templeton. Energy for Sustainable Development , **15** (2011)
2. ISAT/GTZ. Deutsche Gesellschaft für Technische Zusammenarbeit (GTZ),(1999)
3. F. Cotana, A. Petrozzi, G. Cavalaglio, V. Coccia, A.L.Pisello, E. Bonamente, ,Energy Procedia, **61**(2014)
4. P. Raman, V.V.R. Rao, V.V.N.Kishore., Biological Wastes1989,**30**,4 (1989)
5. R. Jyothilakshmi, S.V. Prakash. Procedia Environmental Sciences, 35(2016)
6. V.CN. Nguyen, T.H. Phan,, Vo, H N. Viet. Env, **2**(1) 2012
7. B. Rao, A Maneb A.B. Rao, V. Sardeshpande. Energy Procedia, **54**(2014)
8. M. Collotta G. Tomasoni. Energy Procedia,**128**(2017)
9. T.G.T. Nindhia. I M. Sucipta, I W. Surata, I K. Adiatmika, D.N.K.P. Negara, K.M.T. Negara., International Journal of Renewable Energy Research, .**3**(1)(2013)
10. T.G.T Nindhia, I W. Surata, I D.G.P. Swastika, , P. Widiana. Key Engineering Materials,**705**(2016)
11. T.G.T. Nindhia, I W. Surata , A. Wardana, Materials Science and Engineering, **201** (012021)(2017)
12. I. S. Horváth, M. Tabatabaei, K. Karimi, R. Kumar, Biofuel Research Journal, **10** (2016)
13. I. J. Dioha, C.H. Ikeme, T. Nafi'u, N. I. Soba and M.B.S. Yusuf. International Research Journal of Natural Sciences, **1**(3)2013
14. I W. Surata, T.G.T. Nindhia, I K.A. Atmika, D.N.K. Negara, I W.E.Putra. Energy Procedia, **52**(2014)
15. D. Raha, P. Mahanta, M.L. Clarke. Energy Policy, **68**(2014)

Gas-to-electricity investment planning for power system stability improvement and environmental sustainability in Nigeria

Oludamilare B. Adewuyi[1,*], *Mark. K.* Kiptoo[1], *Ayodeji F.*Afolayan[2], *Ayooluwa .P.* Adeagbo[3] and *Tomonobu* Senjyu[1]

[1]Faculty of Engineering, University of the Ryukyus, 1 Senbaru, Nishihara-cho, Nakagami, Okinawa, 903-0213, Japan
[2]Institute for Sustainable Development, First Technical University, Ibadan, Oyo, 200253, Nigeria
[3]Faculty of Engineering, Adeleke University, Ede, Osun, 232101, Nigeria

Abstract. Power outage is a prominent feature of the current Nigerian power system. However, a properly planned energy sector can help the nations quest for energy sustainability and economic development. Techno-economic assessment of the Nigerian energy facilities for efficient gas-to-grid power integration is presented in this paper using the particle swarm optimization algorithm (PSO) for solving a voltage stability-constrained optimal power flow model on Matlab environment. Investment in gas-fired DG technology can be an economic and sustainable approach for reducing the detrimental effeccts of gas-flaring practices of the petroleum industries on the environment. The technical benefits such as voltage profile improvement and voltage stability enhancement are the main focus of the technical analysis carried out in this study.

1 Introduction

Voltage stability issues and the extreme problems of indiscriminate voltage collapse remain a crucial challenge to the reliable and economic operation of power systems in Nigeria [1]. According to the Transmission Company of Nigeria, the existing generating capacity of the national grid currently stands at 12,522 MW; out of which the average available daily generation is just about 4,000 MW [2]. Only about 59.3 % of the entire population has access to electricity; and the actual national electricity demand is currently above 20,000 MW and still increasing [3]. On the average, more than 20 occurrences of system collapse, both partial and total, is reported yearly in Nigeria [4]. The reason for this has been attributed to the grossly insufficient generation and inadequate transmission network [5]. As it currently stands, the main threat to achieving the Sustainable Development Goals in Nigeria, like in other developing nations, is the problem of poor economy which is principally due to the erratic nature of the energy industry. However, gas-fired distributed generation (DG) technology is helping to change the power system situation in many parts of the world due to cheap source of primary fuel and closeness to the loads. Properly designed DG systems can reduce the risk of stressing the already overloaded transmission lines and improved the system voltage stability [6].

Gas flaring is a common practice that has contributed hugely to both economic losses and environmental degradation within Nigeria. Gas flaring involves the combustion of gas (including useful natural gas) in open flame at oil production sites with no useful mechanism for harvesting useful energy. The economic and environmental impacts assessment of gas flaring has shown that about 47 percent of the total gas produced and 18 percent of the total produced natural gas in Nigeria is flared yearly [7]. An average of about 2.5 billion cubic feet of gas from crude oil production is reportedly being wasted via gas flaring per year in Nigeria [8]. This amounts to credible economic and environmental resources wastage, and has significantly affected the quality of living in terms atmosphere, land and water pollution. However, sufficient amount of useful energy can be produced from gases if proper investment is made on modern and highly efficient natural gas-based energy conversion technologies [9]. By investing in the recent state-of-the-art gas-driven energy technologies and the development of necessary human resources and passing of the petroleum industry bill (PIB), which enables a competitive market for gas sales and ensures better return on investments made on gas infrastructures, Nigeria can rapidly be on her way to economic growth and sustainable development realization.

The renewable technology is still relatively immature in Nigeria, as it is in most of the sub-Saharan African region, due to several reasons ranging from lack of adequate technical know-how to finding suitable sites and obtaining land right-of-way; as well as huge investment financial requirement [10]. However, In [11], it is projected that up to 5000 MW of electricity could be produced daily if the flared estimated 700 million SCF of gas is properly harnessed. A technical and economic analysis of flared gas to electricity potential in Nigeria can help increase the daily direct electricity production by 7500 MW [9]. The transportation Sector, just like the

* Corresponding author: adewuyiobode@gmail.com

power sector, can also experience massive transformation by the introduction of efficient natural gas reclamation and gas-to-energy conversion technologies using clean-burning and efficient natural gas-driven vehicles in Nigeria [12]. The lifecycle CO_2 emission content of natural gas as primary fuel compared to other energy sources and the possible utilization options for clean natural gases for Nigerian situation is illustrated in Table 1 and Figure 1.

Table 1. Lifecycle CO_2 emission intensity [13].

Fuel source	CO_2 emmission (tonnes/GWh)
Coal	888
Oil	733
Natural Gas	**499**
Biomass	450

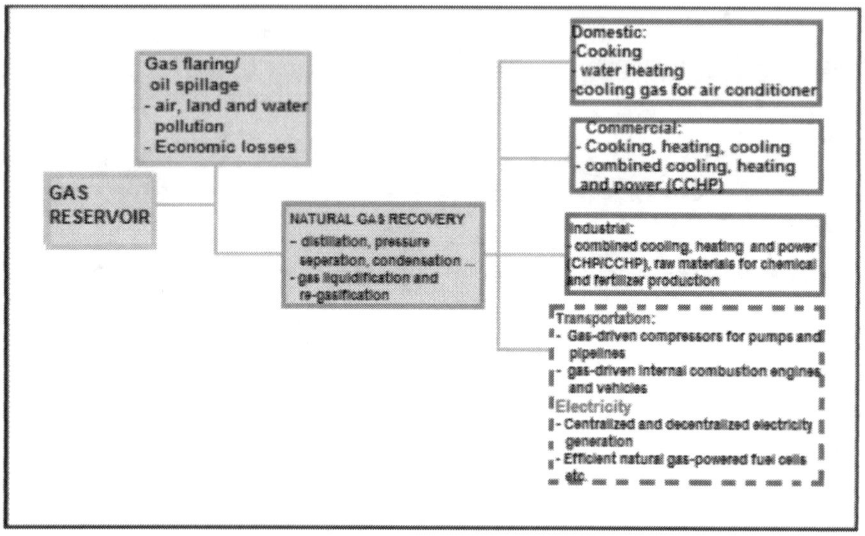

Fig. 1. Possible ways of utilizing natural gas resources for energy production in Nigeria.

2 Voltage stability and DG injection planning

The study of the problems (technical and otherwise) that threatens the secure and reliable operation of power systems under different conditions is known as power system stability analysis. Voltage stability problem, also known as Load stability, is a prominent dynamic power system stability issue that is mainly due to the interaction of the power system loading and insufficient generation and reactive power transfer capacity of existing grid infrastructures [14]. Severe voltage instability can appear in a part or the entire power system due to gross inadequacy of the existing generation and transmission facilities. Hence, monitoring the voltage stability conditions of grid networks is crucial for effective planning of additional power injection from DGs. Several major power system's failure (black outs) in recent times are traceable to the problem of voltage collapse. Hence, while planning for increase generation to meet the growing electricity needs, there is a huge need to give adequate consideration to the capacity and condition of the existing grid facility. One important issue to be considered when adding DG units to power systems is the effects it will have on the power flow and voltage stability margin of the power system. This can be observed by monitoring the nearby buses and transmission lines at the point of DG power injection.

2.1. Effect of DG on voltage stability Margin

The voltage stability margin is usually estimated as the maximum amount of load increase that a power system can bear without violating the stability condition. It is a measure of the shortest distance from current operating point to the point of voltage collapse as illustrated on Figure 2:

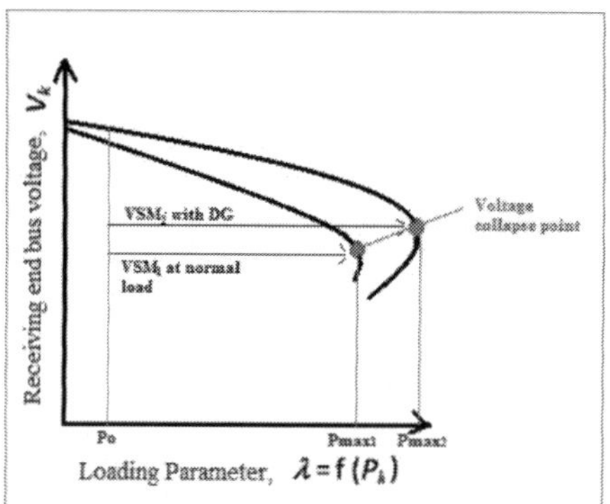

Fig. 2. Power system voltage stability margin

Considering Figure 3, at any point of DG injection, the net effective demand using the constant PQ load model is :

$$P_k + jQ_k = (P_d - P_{dg}) + j(Q_d - Q_{dg}) \quad (1)$$

$$(P_k + jQ_k) = (V_k\angle\delta_k)\frac{(V_i\angle\delta_i - V_k\angle\delta_k)^*}{r_{ik} + jX_{ik}} \quad (2)$$

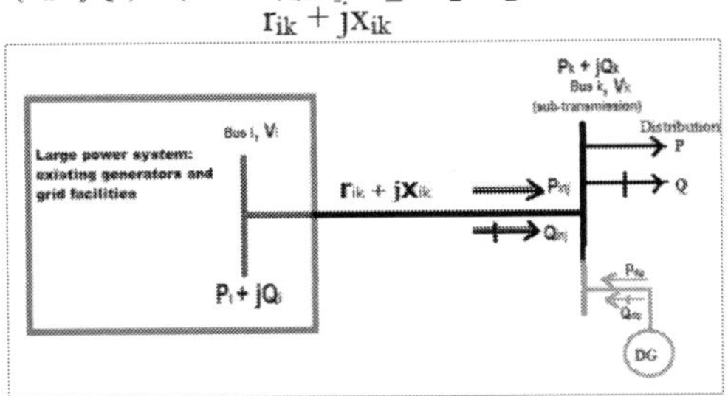

Fig. 3: Illustration of DG injection at a sub-station

Resolving the above equation (2) and separating the real and reactive power components yields:

$$P_k r_{ik} + Q_k x_{ik} + V_k^2 = V_i V_k \cos(\delta_i - \delta_k) \quad (3)$$

$$P_k x_{ik} + Q_k r_{ik} = -V_i V_k \sin(\delta_i - \delta_k) \quad (4)$$

The voltage stability condition of a power system is obtained from algebraic solution of the Eqns. 3 and 4 as [15]:

$$\sqrt{\left[\left(r_{ik}P_k + x_{ik}Q_k \,?\, 0.5V_i^2\right)^2 - \left(r_{ik}^2 + x_{ik}^2\right)\left(P_k^2 + Q_k^2\right)\right]} = 0 \quad (5)$$

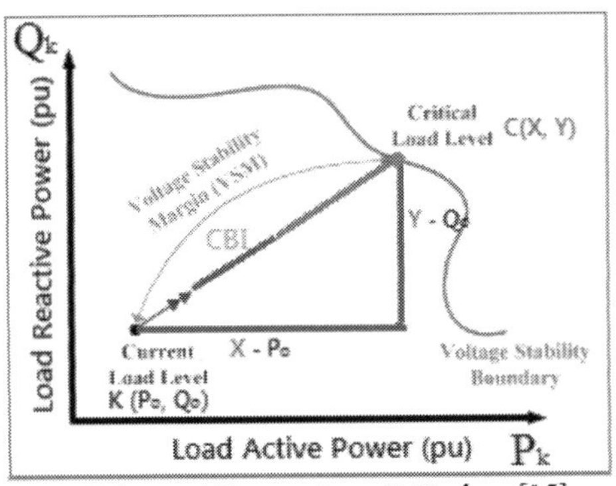

Fig. 4: Voltage stability margin on (P-Q) plane [15].

Figure 4 is a two-dimensional illustration of the real and reactive power loading at the receiving end bus considering a particular bus voltage value. The voltage stability margin, which is the nearest distance of a current operating point, $K(P_o,Q_o)$ to a point, $C(X,Y)$ on critical stability boundary is obtained using the Lagrange multiplier approach as given below :

$$F(X,Y,\lambda) = \sqrt{(X-P_o)^2 + (Y-Q_o)^2} + \lambda\left[\left(r_{ik}X + x_{ik}Y - 0.5V_i^2\right)^2 - \left(r_{ik}^2 + x_{ik}^2\right)\left(X^2 + Y^2\right)\right] \quad (6)$$

The set of equations obtained from the partial derivatives of equation (6) w.r.t. (X,Y,λ) are solve simultaneously to obtain the maximum (critical) real and reactive power loading (X,Y), and the corresponding VSM is obtained according to equations (7)-(9) ;

$$\Delta P = X - P_o \quad (7)$$

$$\Delta Q = Y - Q_o \quad (8)$$

$$CBI = \sqrt{(\Delta P)^2 + (\Delta Q)^2} \quad (9)$$

The closer the CBI value of any of the transmission lines to zero, the closer the power system to voltage collapse.

2.2 Voltage stability index

In a more conventional approach, the assessment of power system voltage stability has been achieved by using several mathematical models collectively known as the voltage stability indices [14]. Some of these voltage stability indices are FVSI, NLSI and LQP and their mathematical expressions are given in reference [16]. These tools are used for planning effective enhancement of power system steady-state performances, based on real or/and reactive power optimal rescheduling while considering DG action and other power system enhancement devices. The system is highly unstable and prone to voltage collapse occurrence if the values of NLSI, FVSI and LQP approach 1.0.

3 Problem Formulation : Gas and electricity optimal power flow (GEOPF)

3.1 Objective function

The mathematical formulation of GEOPF adopted in this work is expressed as a minimization of the total cost of operating the gas-to-grid network infrastructures [17] :

$$\underset{x}{\textbf{Minimize}}\ C_{total}(x) = C_{HT} + C_{CGT} + C_{Grid} + C_{Gpipe} - C_{gs} \quad (10)$$

where $C_{total}(x)$ is the net cost which consists of the following constituent costs: the conventional hydro-thermal generator operating cost C_{HT}, the production and Investment cost on gas turbine C_{CGT}, cost of grid enhancement based on additional power injection C_{Grid}, cost of gas pipeline maintenance C_{Gpipe} and the environmental benefit/sustainability cost of gas (flaring reduction) C_{gs} ; as expressed below:

$$C_{HT} = \sum_{i=1, i\neq G}^{ng} \left(a_i P_{gi}^2 + b_i P_{gi} + c_i\right) \quad (11)$$

$$C_{CGT} = \sum_{i=1}^{G} (P_{Gi} \times K_G) \quad (12)$$

$$C_{Grid} = \sum_{i=1}^{G} (P_{Gi} \times K_{grid}) \quad (13)$$

$$C_{Gpipe} = \sum_{i=1}^{Npipe} (L_{Gpipe} \times K_{Gpipe}) \quad (14)$$

$$Cgs = \sum_{i=1}^{ngs} (K_i^{net} \times Q_{gsi} \times GHV \times 10^6) \quad (15)$$

ng is the total number of existing hydro-thermal generators, G is the number of selected injection point/combined cycle generating plants, $Npipe$ is the number of pipelines and ngs is the total number of natural gas source points. a, b and c are the quadratic fuel cost coefficients and Pg is the active power generated by each unit of the existing generators. K_G and K_{grid} are the corresponding cost per MW of gas plant investment and grid enhancement, respectively. A linear cost model based on the average distance from the oil producing zones to the industrial and commercial zones. k^{net} is the net cost of gas at each source node in \$/MMBTU. Q_{gs} is the optimal amount of natural gas consumed by the generator in CF and GHV is the gas gross heating value in BTU/CF. (BTU means British thermal unit and CF means cubic foot). The cost k^{net} consists of the actual cost of selling the power from the gas to the grid plus the cost of social benefits in terms of reduced environmental pollution less the cost of selling the gas by the gas suppliers. Hence, the total cost of gas C_{gs} is treated as a positive social benefit which is to be maximized.

3.2 System constraints

3.2.1. Power balance and system limit constraints

The system generator, bus and line constraints are given in [10]. The negative PQ load model for DG injection analysis is considered for the power balance equation in the load flow analysis [18].

3.2.2. Voltage stability margin constraint

To keep the power system secure and within the stability limits, the additional constraints based on the critical stability margin (CBI) is considered as given:

$$CBI_i \geq 0.3; i = 1, 2, 3 \ldots nl \quad (16)$$

nl is the number of transmission lines.

3.3 Particle Swarm Optimization (PSO) Algorithm with time-varying acceleration coefficient

The PSO algorithm based on the social behavior and interaction of particles within a swarm is adopted for solving the non-linear optimization problem described in this work; the detail of PSO algorithm is given in [19]. The social and cognitive factors are taken to be varying instead of being constant using the time varying acceleration coefficent (TVAC-PSO) approach described in [20]. Figure 5 is a flowchart that shows the TVAC-PSO algorithm. The PSO and hydrothermal generator parameters are as given in [10].

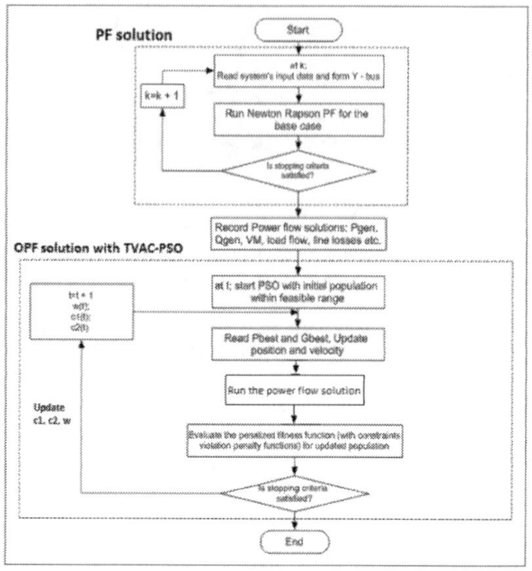

Fig. 5: TVAC-PSO flow chart [20]

The natural gas parameters, using a conversion rate of 3412.14 CF of natural gas = 1 MWh of electricity and 1.00 USD = 360 Naira, are given in Table 2.

Table 2. Natural gas data [7,9].

Symbol	Value (Unit)
K_G	1015.00 (\$/kW)
K_{grid}	3.50 (\$/kW)
K_{Gpipe}	1.80 (\$/kW)
K^{net}	5.50 (\$/CF)
GHV	1150.00 (BTU/CF)

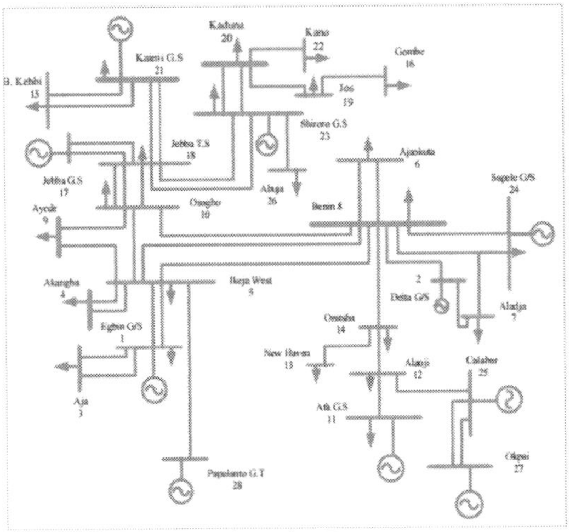

Fig. 6: Nigerian 28 bus Test system

4 Result and discussion

The real Nigerian grid, given in Figure 6, at overloaded operating condition is used for the analysis presented in this study. The system has 28 buses and 52 transmission lines. The voltage stability analysis using CBI parameters are used to select DG power injection points, as indicated in Figure 7. Two conditions were considered in selecting the candidate buses; the optimal voltage stability margin of the transmission line described in equation (5)-(9) and the real power loss sensitivity factor, given by equation (16) as obtained from the exact loss formula [18].

$$L_{fi} = \frac{\partial P_{ik}^{loss}}{\partial P_i} = 2\sum_{k=1}^{N_b}\left[\alpha_{ik}(P_k + \frac{Q_k}{\wp}) - \beta_{ik}(Q_k - \wp P_k)\right] \quad (16)$$

Where;

$$\alpha_{ik} = \frac{r_{ik}}{V_i V_k}cos(\delta_i - \delta_k);$$

$$\beta_{ik} = \frac{r_{ik}}{V_i V_k}sin(\delta_i - \delta_k) \quad \text{and} \quad \wp = \tan(\cos^{-1}(pf))$$

The receiving end buses of the lines with sufficient voltage stability margin (based on real power loadability) and low loss sensitivity are selected. Figure 8 shows the convergence characteristics using TVAC-PSO with optimal total cost of US $1.285 x 10^7 with a gas consumption of 6.312 x 10^6 SCF (*i.e.* Q_{gs}) and injected DG power of 477.45 MW (*i.e.* P_G). There is a credible improvement in the voltage profile with the placement of DGs as it can be seen on Figure 9. However, there is a need for voltage compensation at the buses 16 and 22 to stay within the set voltage magnitude threshold of 0.95 p.u. The injected power from the gas-fired DGs has plausible improving effects on stability margin of most of the lines, especially the real power stability margin as seen in Figure 10. Also, there is an increase in the optimal stability margin of most of the ill-conditioned lines while all the healthier lines, such as lines 1, 2, 3, 4, 14 *etc.* remains within the tolerant stability margin.

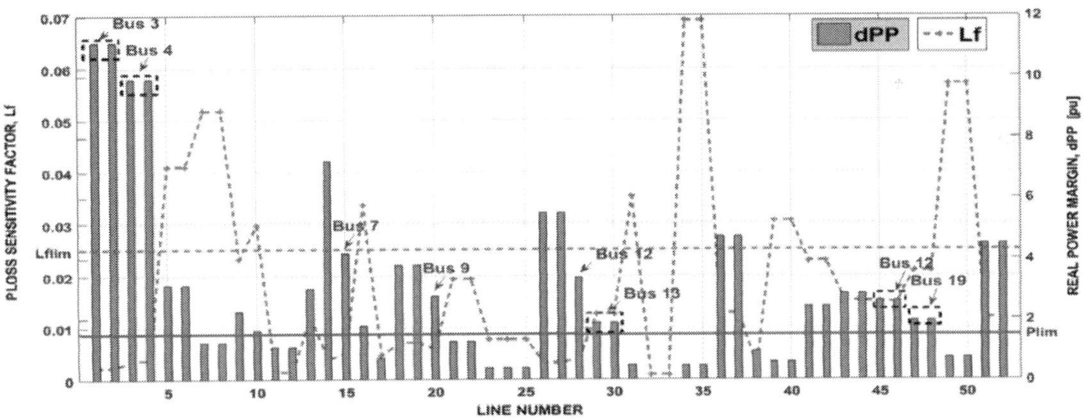

Fig. 7: Identified DG location point

 https://doi.org/10.1051/e3sconf/201912002005

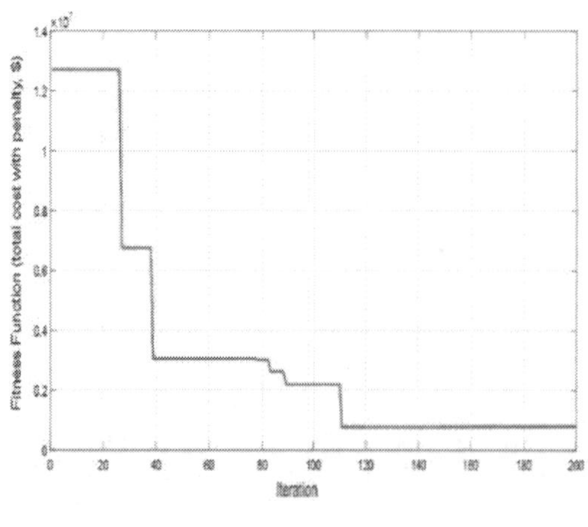

Fig. 8: Convergence characteristics

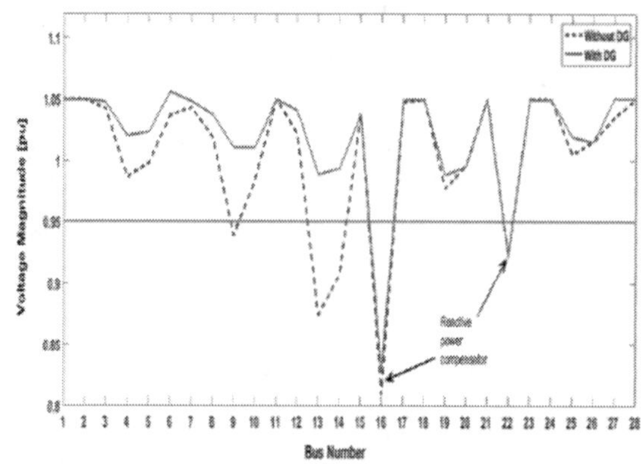

Fig. 9: Bus voltage magnitude

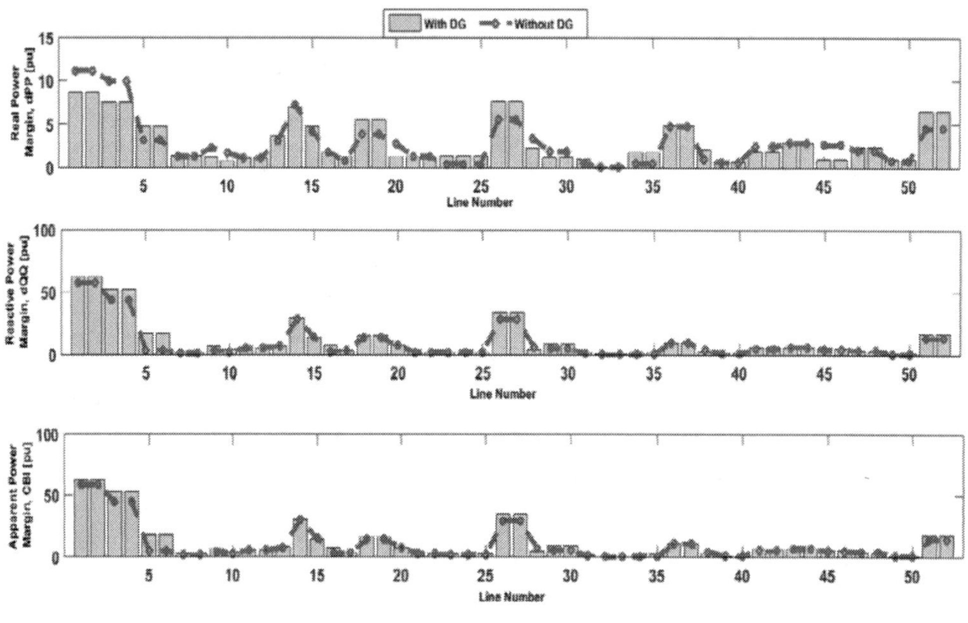

Fig. 10: Effects on optimal stability margin

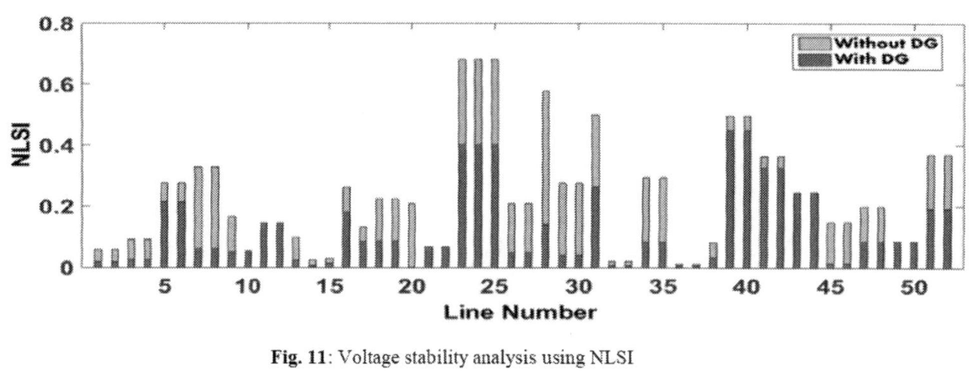

Fig. 11: Voltage stability analysis using NLSI

The overall influence on the voltage stability condition of the power system was analyzed using the NLSI (equation 17) and presented in Figure 11.

$$NLSI = 4\left(\frac{P_k r_{ik} + Q_k x_{ik}}{V_i^2 cos_\delta^2}\right) \quad (17)$$

The stability condition of the power system is improved as reflected on the reduction in the NLSI value for almost all the transmission lines. For lines 23, 24 and 25, which are the most critical lines, there is a reduction from 0.7000 to about 0.4023. This improves the security of the power system and keep it farther away from voltage collapse.

Universal access to clean and reliable energy, while keeping the environment pollution-free is a key component of the United Nations' Sustainable Development Goals. Hence, there is currently great interest around the world on the adoption of alternative energy sources, both renewable and non-renewable (clean burning), towards achieving energy and environment sustainability. A techno-economic approach to achieve some of these targets at minimal cost and with positive impact on the immediate environment has been investigated and analyzed in this work, using the Nigerian energy sector situation. The optimization goal is to ensure a voltage-secured use of the existing grid facility and the sustainability goal is to demonstrate how much social, technical and economic benefits can be derived from the judicious use of available energy resources and infrastructures, especially for an oil-producing and developing nation like Nigeria.

5 Conclusion

References

1. A.O. Ibe, E. Omorogiuwa and I. Kayode, "voltage stability analysis of nigerian power system using distributed generation," Applied Research Journal, vol. 3, issue 2, pp. 53–62, 2017.
2. A. E. Israel, A. A. Enesi, and A. J. Kufre, "Assessment of Nigeria's power situation and the way forward," *International Journal of Latest Research in Engineering and Technology*, vol. 1, issue 34, pp. 24–31, 2015.
3. Sulaimon Salau, "Nigeria's power demand to hit 88,282 MW by 2020," The Guardian, https://guardian.ng/business-services/nigerias-power-demand-to-hit-88282mw-by-2020/, 2016 [url date: 2016-06-28]
4. U. C. Ogbuefi, C. L. Ugwu and N.O. Ogbogu, "Analysis of Nigeria power system voltage collapse incidences from 2000 to 2017 *IOSR Journal of Electrical and Electronics Engineering*, vol. 13, number 2, pp. 28-34, 2018.
5. B. O. Akinloye, P. O. Oshevire and A. M. Epemu, "Evaluation of system collapse incidences on the Nigeria power system," Journal of Multidisciplinary Engineering Science and Technology, vol. 3, number 1, pp. 3707-3711, 2016.
6. Sudipta Ghosh, S. P. Ghoshal and Saradindu Ghosh, "Optimal Sizing and Placement of Distributed Generation in a Network System," International Journal of Electrical Power and Energy Systems, vol. 32, 2010, pp. 849–856.
7. O. Adekomaya, T. Jamiru, R. Sadiku, Z. Huan and M. Sulaiman, "Gas flaring and its impact on electricity generation in Nigeria", Journal of Natural Gas Science and Engineering, vol. 29, pp. 1-6, 2016.
8. U. Udok and E. B. Akpan, "Gas flaring in Nigeria: problem and prospects," *Global Journal of Politics and Law Research*, vol. 5, issue 1, pp. 16–28, 2017
9. E. N. Ojijiagwo, C. F. Oduoza, N. Emekwuru., "Technological and economic evaluation of conversion of potential flare gas to electricity in Nigeria". *Procedia Manufacturing* 17, 444–451, 2018.
10. O.B. Adewuyi, M.E. Lotfy, B.O. Akinloye, H.O.R. Howlader, T. Senjyu and K. Narayanan, "Security-constrained optimal utility-scale solar PV investment planning for weak grids: Short reviews and techno-economic analysis", *Applied Energy*, Vol. 245, pp. 16–30,2019.
11. M. Eboh, "Nigeria losing N 868m daily to gas flaring," https://www.vanguardngr.com/2018/03/nigeria-losing-n868m-daily-gas-flaring-nnpc/, [date accessed: 2019-02-08].
12. S. O. Giwa, C. N. Nwaokocha, B.O. Odufuwa, "Mitigating gas flare and emission footprints via the implementation of natural gas vehicles in Nigeria". *Energy Policy* 111, 193–203, DOI: https://doi.org/10.1016/j.enpol.2017.09.027 , 2017.
13. WNA, Comparison of lifecycle greenhouse gas emissions of various electricity generation sources, 2011. http://www.world-nuclear.org/uploadedFiles/org/WNA/Publications/Working_Group_Reports/comparison_of_lifecycle.pdf
14. O. B. Adewuyi, M. S. S. Danish, A. M. Howlader, T. Senjyu and M. E. Lotfy, "Network structure-based critical bus identification for power system considering line voltage stability margin," *Journal of Power and Energy Engineering*, vol. 6, pp. 97–111, 2018.
15. M. Furukakoi, O. B. Adewuyi, M. S. S. Danish, A. M. Howlader, T. Senjyu and T. Funabashi, "Critical

Boundary Index (CBI) Based on Active and Reactive Power Deviations," *International Journal of Electrical Power and Energy Systems*, vol. 100, issue 1, pp. 50–57, 2018.

16. M. S. S. Danish, A. Yona and T. Senjyu, "A Review of Voltage Stability Assessment Techniques with an Improved Voltage Stability Indicator," *International Journal of Emerging Electric Power Systems*, vol. 16, issue 2, pp. 107–115, 2015.
17. S. An, "Natural Gas and Electricity Optimal Power Flow", Ph.D. Thesis, Oklahoma State University, USA, 2004.
18. A. Tah. and D. Das: "Novel Analytical Method for the Placement and Sizing of Distributed Generation unit on Distribution Networks with and without considering P and PQV buses", *International Journal of Electrical Power and Energy Systems*, vol. 78, pp. 401–413, 2016
19. J. Kennedy and R. Eberhart: "Particle Swarm Optimization", *IEEE International Conference on Neural Networks*, vol. 4 pp. 1942 – 1948, 1995
20. A. Ratnaweera A., S. K. Halgamuge and H. C. Watson, "Self-Organizing Hierarchical Particle Swarm Optimizer with Time-Varying Acceleration Coefficients", *IEEE Transactions on Evolutionary Computation*, Vol. 8, Issue 3, pp. 240 – 255, 2004.

Integrated ecosystems management of the coastal zones in the Artic zone of the Russian Federation: problems and solutions

Alexey Konovalov, *Galina* Baturova, *Tatyana* Gavrilenko, and *Olga* Grigorenko*

MIREA – Russian Technological University, 119454 Moscow, Russia

Abstract. The article is devoted to development, implementation and information support of integrated approach to ecosystem management in the Arctic zone of the Russian Federation in the context of global climate change. According to the adopted strategic planning documents for the development of the Arctic zone of the Russian Federation, integrated ecosystems management of the coastal zones is an issue that must be dealt with expeditiously. Solution of this problem is an urgent task in the state policy of the Russian Federation. The article presents the methodology for development, implementation and information support of integrated approach to ecosystem management in the Arctic zone of the Russian Federation as well.

1 Introduction

1.1 Specifics of coastal zones in the Arctic zone of the Russian Federation

The importance of Arctic spaces and resources in the life of mankind and the formation of the global gross product has increased dramatically. It is predicted that as a result of global climate change the dominant position in the structure of world trade can reach the commodity flows passing through the high-latitude transport and communication highways of the Arctic region. Despite all the technical difficulties of sailing in the High North geographically the North Sea Route (NSR) is the shortest route connecting Europe with the Far East, Asia and the Western part of North America (Figure 1). The use of the NSR can reduce the length of the reference route between Hamburg and Yokohama 2.3 times when sailing around Africa and 1.4 times in comparison with the passage through the Suez Canal.

For example, we developed two scenarios for cargo transportation volume along the NSR: baseline and target. The baseline scenario extrapolates the available data. The second scenario takes into account the possible global climate change influence. According to these scenarios, the expected cargo transportation volume along the NSR is about from 76.5 million tons (baseline scenario) up to 109.7 million tons (target scenario) (Figure 2).

The full-scale development of mineral and energy resources of the richest Arctic continental shelf directly affects the structure of the world's energy supply. The usage of renewable energy sources in the Arctic region including wind, geothermal, circulating ocean, wave, planetary gravity (tides) energy generates a wide and huge range of technological innovations used in various spheres of life. According to the Food and Agriculture Organization scenario forecasts, there will be a significant growth of marine industrial fisheries in the nearest future. The Arctic region plays an important role in this sphere. The global climate-forming function of the Arctic Ocean catalyzes the intensification of fundamental marine studies. All these trends led to the necessity of systemic evaluation of the impact of maritime activities on the environment, as well as its control and transition from sectorial to integrated management [1].

As a result of the intensification of maritime activity the competition and conflicts for the same areas as well as resources of the Arctic Ocean occur often between branches of marine complex. The occurring competition demand accommodation of conflicting interests by means of integrated coastal zone management (ICZM) and marine spatial planning (MSP). Goals of such planning are based on provision of a balance between economic activity, implementation of social policy priorities and environment-oriented tasks. The MSP instruments (including functional zoning of marine areas according to biological productivity and biodiversity; evaluation of their integral vulnerability to anthropogenic impacts; use of geographic information systems and technologies; determination of high conflict zones and their identification according to the compatibility/incompatibility of types of marine activities; establishment of "green" corridors for migration of biological resources; efficient supra-departmental control of implementation of the MSP programs and their regulatory support, etc.) allow to increase significantly the level of protection of marine ecosystems from degradation, reduce conflicts between branches of marine complex.

The Strategy for the Maritime Activities Development of the Russian Federation until 2030 was approved by the Government of the Russian Federation in 2010 [2]. According to the Strategy one of the strategic tasks is devoted to elaboration and implementation of the

* Corresponding author: bagalina@bk.ru

programs for coastal territories and marine areas integrated development as a component of strategies and programs for social and economic development of coastal regions of the Russian Federation. The Strategy for the development of the Arctic zone of the Russian Federation, approved by the President of the Russian Federation [3], among the priorities and main activities put an increased focus on elaboration and approbation of the integrated coastal zone management models in the Arctic zone of the Russian Federation (AZRF). Paragraph 31 of the Action Plan for implementation of the Strategy on social and economic development of the Ural Federal District for the period until 2020 adopted by the Government of the Russian Federation in 2012 [4] sets a task to elaborate the program for Yamal-Nenets autonomous district's coastal territories and marine areas integrated development. Solution of this task should be closely related to the issues of MSP as a tool for spatial reflection of the strategic decisions approved in programs for integrated coastal zone management / coastal territories and marine areas integrated development (ICZM/CTMAID).

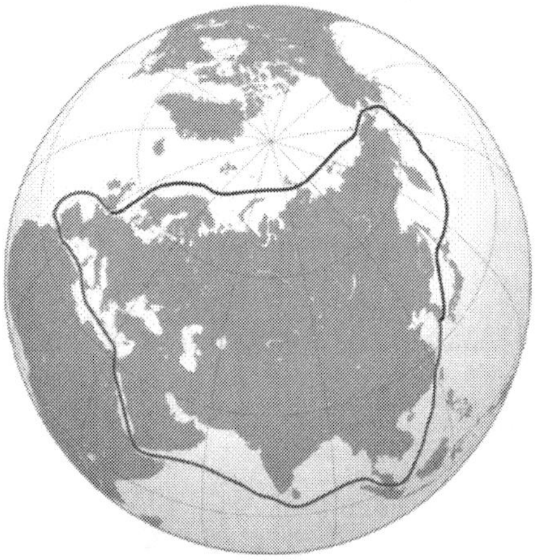

— by the Northern Sea Route

— by alternative south route

Fig. 1. "Asia-Europe" passage

The strategic planning documents emphasize that economic development in the AZRF are closely related to the Arctic Ocean, maritime transportation via the NSR. Most of the settlements here are located on the coast of the Arctic Ocean [5]. Global climate change, a shift in economic activity in the shelf zone of the Arctic seas leads to an increasing role of the marine factor in the economic and social development of the AZRF [6]. The coastal location of the macroregion differentiate it from the "continental" parts of the North of Russia and determines the uniqueness of the regional economic complexes formed here [7].

In comparison with other Arctic States, it is in the AZRF that the most powerful industrial layer has been created. Here the part of value added of extractive industries is about 60%, while in Alaska and Arctic Canada it does not exceed 30%, and in the countries of Fenno-Scandia (including the Northern regions of Norway, Sweden, Finland, as well as Greenland and Iceland) it reaches only 15%. At the same time, the fragility and vulnerability of the Arctic environment demand the development of the environmentally friendly technologies that do not cause irreversible environmental damage during the exploitation of the rich resources of the AZRF.

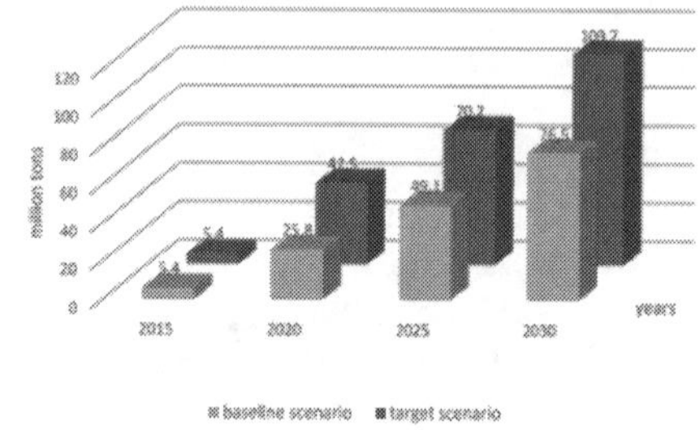

Fig 2. Forecast of cargo transportation volume along the NSR

The Arctic region is undergoing pronounced climatic changes, which affect all components of the environment, including aquatic ecosystems, the water balance of river basins, cryogenic processes, etc. [8]. The state and development of various types of infrastructure in the Arctic are subject to natural risks associated with adverse hydrological processes: floods, congestion, and blockages, ice formation, etc. One of the main factors affecting the processes of flow formation is permafrost [9]. During the recession of permafrost, water exchange increases between surface and groundwater, groundwater flow, including the movement of water in soils, supra-permafrost, and sub-permafrost. Direct and indirect evidence of such changes is observed in various cold regions of the world [10, 11]. Different types of models are applied to the modeling of the influence of hydrometeorological, hydrological, cryolithological, ecological and other parameters on certain types of infrastructure in the Arctic [12-16].

1.2 International practices of integrated approach implementation to regulation of maritime activities

International practices of integrated approach implementation to regulation of maritime activities show, that various countries have started to use MSP to achieve sustainable use, including the goal of developing a "Blue Economy". About 70 countries now have MSP initiatives.

For example, in 2005 in the Netherlands there was a methodology on MSP of Dutch sector of North Sea developed, which was determined by economic growth and advance in business activity (wind power engineering development, extraction of commercial minerals, mariculture, tourists flow increase, necessity for organizing new specially protected natural areas). Principles and mechanisms for the implementation of

MSP are fixed in Integrated Management Plan for the North Sea. The main aims include creation of environmentally safe and stable marine environment, attraction of additional investments and execution of maricultural business-projects, formation of efficient spatial management tool, which will allow to specify zones with a definite kind of use, etc. Among the general tasks are: 1) use of cartographic documents to define aquatorium sections, where activities are already being carried out, and to find out prospective zones to become utilized in future; 2) environmental monitoring; 3) integral assessment for the purpose of issuance of licenses, permits; 4) integrated scientific researches of marine environment; 5) jurisdictional disputes resolution in cases when one aquatorium user's activities do harm to another one; 6) launching the different economic operators' initiative to share aquatorium. In 2015 the Netherlands have published North Sea 2050 Spatial Agenda as a legal platform for using MSP.

Belgium was one of the first European countries to elaborate and implement MSP documents in a very short time. The main cause for the country to get into forming MSP system was development of wind power engineering in offshore water and international commitments on protection of bioresources and ecosystems. MSP in Belgium pursues economic and ecological aims – development of wind power engineering on the open sea, demarcation of specially protected areas, implementation of plan of sand and gravel extraction, conservation of maritime biodiversity, management of different kinds of economic activity, influencing on maritime ecosystems. Belgian MSP scheme was worked out in 2003. It is estimated that in prospect MSP documents will preserve their multipurpose nature and will include additional zones of nautical archaeology objects protection, zones for studying alternative methods of fishing, etc. As part of the Plan of North Sea aquatorium development, they developed the method of integrated use of offshore zone, with special emphasis on alternative use scenario. Scenarios represent "views on projects which are supposed to be implemented in the nearest future in North Sea areas". A revised Marine Spatial Plan (2020-2026) has been drafted in 2017.

MSP in Germany rests upon the Federal Law on Land Use Planning, application of which was extended over water areas, and is based on the principles of economic empowerment via orderly spatial development and optimization of maritime spaces management, protection of maritime ecosystem against pollution. In 2007 Federal Sea and Hydrographic Agency of Germany prepared the project of MSP of the country's exclusive economic zone in the North Sea and the Baltic Sea. Federal marine spatial plans approved for Baltic and North Seas exclusive economic zone (EEZ) in 2009.

In April 2006 the Government of Norway started development of overall plan on managing Norwegian part of Barents Sea, which secures management of all economic activities with a view to ensuring safety of ecosystems' quality and functions. Plan implementation involves achievement of a number of aims:

resolution of conflicts between economic activity and environment protection;

continuation of development of existing managerial measures, regulating different kinds of activities;

prioritized solution of environmental problems;

expansion of international cooperation.

Ecosystem-based MSP plan for Barents Sea was updated in 2011 and 2015, covers Barents Sea part of Norwegian EEZ. Analogous management plans are developed for Norwegian Sea and Norwegian part of North Sea. Ecosystem-based MSP plan for Norwegian Sea was approved in 2009 and updated in 2017, ecosystem-based MSP plan for North Sea was approved in 2013.

Since the beginning of 2000 Australia started development of conception "maritime bioregionalization" as the basis for development of MSP. Bioregionalization is directed towards depiction of spatial structures of ecosystems at scales of corresponding MSP schemes. In 2005 the Government of Australia adopted a program of regional maritime spatial planning, developed on basis of the law on environmental protection and conservation of biodiversity. According to this legislative act, the main principle of MSP documents is protection of maritime spaces.

In 1997, according to national policy of China Government, which was based on "steadfast implementation of laws, regulating management and protection of land, water, forest, fossil resources and seas", State Administration of Oceans suggested to develop the law of controlling maritime areas, which came into effect on the 1 of January 2002. The law warrants that any use of maritime spaces should correspond to the scheme of maritime zoning, approved by the state. After the extensive data gathering, active researches and numerous consultations, the scheme of national maritime zoning was presented to the State council of PRC and approved on 22 August 2002. Its implementation is the basis for creation of regional planning systems and schemes of overall development and preservation of maritime spaces of China. Marine Functional Zoning Plan for EEZ approved in 2012.

One of the well-known example of MSP as the mean of natural ecosystems preservation is national marine preserve Florida Keys in south-east part of the USA. The aims of MSP in Florida Keys imclude protection and preservation of valuable components of the ecosystem, for the region of high conservation value to be able to run its natural course with minimal human impact. There were a number of zones kinds with different protection statuses marked in the preserve: Special Protected Natural Areas; ecological reservations; regions of natural resources use.

There are several international organizations playing integrating role in development and implementation of MSP in Europe:

HELCOM (The Baltic Marine Environment Protection Commission);

VASAB (Vision and Strategies around the Baltic Sea – Russia presides over VASAB in 2012);

ICES (International Council for the Exploration of the Sea);

OSPAR (Convention for the Protection of the Marine Environment of the North-East Atlantic).

Today, geo-information systems (GIS) provide significant assistance for decision-making. For example, the Ministry of the Interior of Japan uses big data analytics to provide timely information on evacuation during disasters. Such a system more effectively notifies residents of evacuation plans during environmental disasters, such as floods and landslides. The system collects statistics on various PALs, as well as user data from social networks. This data allows for providing more accurate and personalized information based on the location of users and the level of danger. During major disasters such as tsunamis and earthquakes, rescuers use geolocation information from smartphones and other mobile devices during the evacuation. The information is disseminated both through traditional notification channels (radio and television), and through the Internet, mobile networks.

According to the adopted strategic planning documents for the development of the Arctic zone of the Russian Federation, integrated ecosystems management of the coastal zones is an issue that must be dealt with expeditiously. Solution of this problem is an urgent task in the state policy of the Russian Federation. It is necessary to development, implementation and information support of integrated approach to ecosystem management in the Arctic zone of the Russian Federation in the context of global climate change as soon as possible.

2 Methodology For Development, Implementation And Information Support Of Integrated Approach To Ecosystem Management In The Arctic Zone Of The Russian Federation In The Context Of Global Climate Change

Technologies for integrated ecosystem management in the AZRF and the creation hardware-software system for monitoring of their implementation in the strategic planning in harsh climate should solve such issues as:

development of technologies for integrated coastal zone management and marine spatial planning in the AZRF, as well as a set of scientific and technical solutions for their implementation in the strategic planning for social and economic development of the macroregion;

database population and updating for the development and implementation of programs for integrated coastal zone management / coastal territories and marine areas integrated development and schemes of maritime spatial planning in the Arctic region;

creation of an information system for monitoring of ecosystem management technologies implementation in the system of strategic planning for long-term infrastructure development in harsh climate;

balancing of spatial development of the Arctic zone of the Russian Federation in the context of global climate change;

comprehensive and advanced development of infrastructure in the Arctic zone of the Russian Federation on the principles of public-private partnership;

technological, organizational and institutional innovation in the economy and social sphere of the Arctic zone of the Russian Federation, development of knowledge intensive services and elements of the service economy of navigation on the Northern sea route;

ecological safety ensuring in the Arctic zone of the Russian Federation, improvement of the quality of life of the population living and working in the macroregion.

Methodologically ICZM/CTMAID program components depicted in the MSP schemes include some parts. Among them:

1) development and implementation of programs for integrated coastal zone management / coastal territories and marine areas integrated development in the Arctic zone of the Russian Federation;

2) priority development of transport, industrial, energy, information and telecommunication [17], service and social infrastructure in the interests of sustainable social and economic growth of the macroregion;

3) development of knowledge intensive services and elements of the service economy of navigation (logistics, hydrometeorology, navigation and hydrographic support, etc.);

4) development of programs to ecological safety ensuring of coastal territories and marine areas in the Arctic zone of the Russian Federation;

5) social and sociocultural design, including projects in education, medicine, culture (by using of remote technologies in education, culture, development of telemedicine), preservation of ethnoforming activities and traditional crafts of indigenous peoples of the North, Siberia and the Far East of the Russian Federation living in the Arctic zone of the Russian Federation;

6) development of strategies and programs for adaptation of regional marine complexes to global climate change;

7) creation of an information system for long-term forecasting of infrastructure development taking into account the impact of natural and climatic factors, including global climate change;

8) formation of regional transport subsystems using high-speed amphibious equipment: hovercrafts, floating amphibians with wheeled or tracked chassis, air cushion platforms, seaplanes, amphibious aircraft with air cushion chassis, etc.;

9) projects of deep and non-waste processing of aquatic bioresources, renewal of ocean fishing, development of mariculture and fishing of non-traditional and underutilized species of hydrobionts, creation of technoparks of marine genetic and biotechnology, enterprises for the production of innovative bioproducts;

10) the use of renewable energy sources of marine origin (tidal power plants, wind farms, spar buoys, wave power plants, floating hydrothermal power plants, osmotic power plants, technologies for the production of biofuels from sea weed, etc.);

11) new technologies (nanotechnology, biotechnology, information, cognitive technology, cryotechnology, CALS-technology, etc.) and materials implementation;

12) development of tourist and recreational complex (cruise, ecological, cultural and educational, pilgrimage, scientific and scientific-expedition, adventure, extreme, business, Congress and intensive tourism (conferences and seminars), sports, medical and General health tourism, diving, etc.);

13) clusterization of spatial development and creation of a network of specialized clusters;

14) preservation of cultural, natural and spiritual heritage;

15) comprehensive studies of the nature of the Arctic ocean;

16) spatial development planning.

These tasks will be focused on supporting decision-making in the interests of the effective infrastructure development of the Arctic zone of the Russian Federation - the country's key macro-region, which plays a significant role in the socio-economic development of Russia.

According to our estimation, development and implementation of integrated approach to ecosystem management in the Arctic zone of the Russian Federation will lead to such social and economic effects, as shown in Figure 3-4.

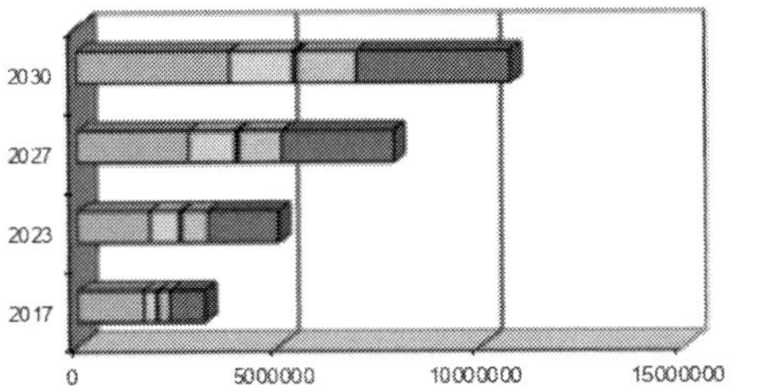

Fig. 3. Effects in economic sphere: volume and structure of the gross regional product (mln. RUB)

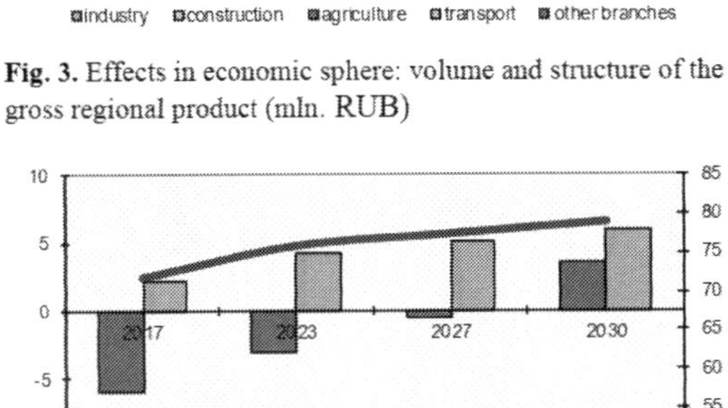

Fig. 4. Effects in social sphere: key demographic indicators

3 Conclusion

1. MSP in Russia could be implemented in the framework of coastal territories and marine areas integrated development program; so, both these elements (CTMAID and MSP) may be an integral part of the Russian state strategic planning system.

2. Guidelines for the elaboration and implementation of the CTMAID programs and MSP, as well as their legislative formalization, should be approved as soon as possible.

3. First of all "coastal territories and marine areas" are the objects of the state regulation, management and control, so when determining its borders, it is desirable not to use only geographical criteria, but the principle connected with the impartibility of the administrative and territorial structure of the Russian Federation.

4. Priority in the CTMAID programs and MSP should be given to social and economic development to meet the "green economy" principles, taking into consideration the valuation of ecosystem services.

5. The length of the Russian coastline leads to localization of maritime industries in poles of growth on perimeter of the maritime borders of the Russian Federation; that's why on first stage CTMAID and MSP should be focused on areas with intense maritime activities and with stakeholders' conflict interests, as well as on special tasks, for example, connected with traditional activities of indigenous peoples of the North, Siberia and Far East of the Russian Federation.

References

1. A.M. Konovalov, G.V. Baturova. *Arctic strategy of Russia* (Saarbrucken: LAP, 2013)
2. Russian President. Decree No. Pr-232. *On Approval of the Strategy for the development of the Arctic zone of the Russian Federation and ensuring national security for the period until 2020* (2013)
3. Russian Government. Decree No. 2205-p. *On Approval of the Strategy for the development of Maritime activities of the Russian Federation until 2030* (2010)
4. Russian Government. *Action Plan for implementation of the Strategy on social and economic development of the Ural Federal District for the period until 2020* (2012)
5. A.M. Konovalov. *Strategic planning of the Arctic zone of the Russian Federation development* (Moscow: SOPS, 2015)
6. G.V. Baturova. *The system-forming role of the Northern sea route in the development of the Arctic zone of the Russian Federation* (Moscow: State Duma Committee on regional policy and problems of North and Far East, 2015)
7. A.M. Konovalov, G.V. Baturova. *Strategic planning of social and economic development of the Arctic zone of the Russian Federation.* Region: economics and sociology **4(88)**, 3-23 (2015)
8. R.M. Holmes, M.T. Coe, G.J. Fiske, T. Gurtovaya, J.W. McClelland, A.I. Shiklomanov, R.G.M. Spencer, S.E. Tank, A.V. Zhulidov. *Climate Change Impacts on the Hydrology and Biogeochemistry of Arctic Rivers.* Climatic Change and Global Warming of Inland Waters: Impacts and Mitigation for Ecosystems and Societies (John Wiley & Sons, Ltd, Chichester, UK, 2012)

9. I. de Grandpré, D. Fortier, E. Stephani. *Degradation of permafrost beneath a road embankment enhanced by heat advected in groundwater.* Canadian Journal of Earth Sciences **49(8)**, 953-962 (2012)
10. L.D. Hinzman, C.J. Deal, A.D. McGuire, S.H. Mernild, I.V. Polyakov, J.E. Walsh. *Trajectory of the Arctic as an integrated system.* Ecological Applications **23(8)**, 1837-1868 (2013)
11. C. Harris, L.U. Arenson, H.H. Christiansen, B. Etzelmüller, R. Frauenfelder, S. Gruber, W. Haeberli, C. Hauck, M. Hölzle, O. Humlum, K. Isakswen, A. Kääb, M.A. Kern-Lütschg, M. Lehning, N. Matsuokai, J.B. Murton, J. Nötzli, M. Phillips, N. Ross, M. Seppälä, S.M. Springman, D.V. Mühll. *Permafrost and climate in Europe: Monitoring and modelling thermal, geomorphological and geotechnical responses*. Earth Science Reviews **92**, 117-171 (2009)
12. T. Reichler, J. Kim. *How Well Do Coupled Models Simulate Today's Climate?* Bulletin of the American Meteorological Society **3**, 303-311 (2008)
13. C.R. De Freitas, E.A. Grigorieva. *Role of acclimatization in weather-related human mortality during the transition seasons of autumn and spring in a termally extreme mid-latitude continental climate.* International Journal of Environmental Research and Public Health **12** (2015)
14. P.G. Dixon, M. Allen, S.N. Gosling, D.M. Hondula, V. Ingole, R. Lucas, J. Vanos. *Perspectives on the Synoptic Climate Classification and its Role.* Interdisciplinary Research Geography Compass volume **10(4)**, 147-164 (2016)
15. H. Li, C. Goh, K. Hung, J.L. Chen. *Relative Climate Index and Its Effect on Seasonal Tourism Demand.* Journal of Travel Research (published online January 23, 2017)
16. H. Li, H. Song, L. Li. *A Dynamic Panel Data Analysis of Climate and Tourism Demand: Additional Evidence.* Journal of Travel Research volume **56(2)**, 158-171 (2017)
17. T. Y. Gavrilenko, I. P. Provorova. Network economy as a phenomenon of the information society. Russian Technological Journal, volume 1 (10), 53-62 (2016)

Study on Stability and Reliability of NOx Ultra-low emission in Coal-fired Power Plants in China

Yang Zhang[1,2*], *Xiaoying* Xin[1], *Da* Hu[1], *Wentao* Zhu[1], *Yue* Zhu[1]

[1]Huadian Electric Power Research Institute Co. Ltd., Hangzhou 310030, China
[2]State Key Laboratory of Clean Energy Utilization, Zhejiang University, Hangzhou 310027, China

Abstract. Along with the rapid progress of flue gas ultra-low emission reconstruction in coal-fired power plants in recent years, the stability and reliability problems in the operation of the project have emerged, among which the denitrification problem is the most acute. Given the present situation of the NOx ultra-low emission in coal-fired power plants, several typical problems about stability and reliability have been analyzed and discussed, including flow fields uniformity, low load denitrification, in-line meter accuracy, catalyst life on the basis of extensive and thorough investigation. Relevant guidance and countermeasures are proposed to provide further reference for the following denitrification projects and operations.

1 Introduction

In recent years, the transformation of ultra-low emission of coal-fired power plants in China has made a rapid progress. With the development of operation and maintenance work after the transformation and the accumulation of operation and maintenance experience, the flue gas pollutant treatment has been significantly improved, and the number and duration of unplanned outage and excessive standard emissions have decreased year by year. However, under the influence of unit load, design margin, standby equipment, operation and maintenance, and other factors, there are still many problems in stability and reliability when meeting the emission requirements[1], especially NOx ultra-low emission operation.

In view of the current status of ultra-low emission of NOx in coal-fired power plants, the widely used SCR denitrification technology was taken as the key research object, extensive and thorough investigation and research on the application of key technologies were conducted. The results show that the main problems in the operation of denitrification unit are low denitrification efficiency, excessive NOx emission, excessive ammonia consumption, excessive ammonia escape concentration, increased air preheater resistance, large pressure difference in the denitrification unit, dust accumulated abrasion of the catalyst, and inoperable at a low load. Given the present situation of the NOx ultra-low emission in coal-fired power plants, several typical problems about stability and reliability have been analyzed and discussed, including flow fields uniformity, low load denitrification, in-line meter accuracy, catalyst life on the basis of extensive and thorough investigation. Relevant guidance and countermeasures are proposed to provide further reference for subsquent denitrification projects and operations.

* Corresponding author: cola1860@163.com

2 Flow Fields Uniformity

The flow field problem is the core problem of the whole denitration plant construction, operation and maintenance. At present, the problems encountered in actual SCR denitrification operation, such as the denitrification efficiency failing to reach the design value, excessive ammonia escape concentration, and ash accumulation and wear of the catalyst, are often caused by uneven distribution of the flow field. Temperature field, velocity field and concentration field, as three key characterization parameters of the flow field, determine the comprehensive performance of the denitrification unit. For example, the uneven temperature field of the same section leads to the uneven denitrification performance of the catalyst. Non-uniform velocity field leads to inconsistent residence time of flue gas, which will cause ash accumulation, clogging and wear in the denitrification reactor, thus affecting the performance of the denitrification device. The uneven concentration field will lead to the deviation of ammonia nitrogen distribution in the denitrification reactor which will bring about excessive ammonia escape and cause damage to the air preheater and other downstream equipment.

For the design of denitrification unit, it is extremely important to take appropriate measures to ensure the uniformity of flue gas velocity and reactant distribution at catalyst inlet section[2]. In the design stage of denitrification engineering, numerical flow field simulation and physical model test studies are generally adopted to ensure the flow field at the catalyst inlet section and the uniformity of ammonia injection mixture in the denitrification reactor through the arrangement of ammonia injection mixing device and the setting of flow

https://doi.org/10.1051/e3sconf/201912003002

deflector[3]. However, due to the limited reconstruction space, compact flue layout, large changes in flue section and many sharp turns, the actual project operation effect is often not ideal. When NOx emission is limited to $50mg/m^3$ as required by the ultra-low emission, the problem of flow field uniformity becomes more prominent[4]. Even if the engineering design is relatively reasonable, the problem of flow field will occur in actual operation due to the change of boiler combustion status, the change of flow field components in denitration reactor or the change of regulating performance of ammonia injection valve. It often leads to the increase of ammonia escape concentration and the obvious aggravation of air preheater blockage. At present, the air preheater resistance of some units has reached to more than 2000Pa, and even some units have to reduce the load operation due to insufficient fan output. Some units with high flue gas temperature even have NH_4HSO_4 sticking on the pole line of the ESP plate and the draft fan blade. As shown in Figure 1, ESP polar sticky dirty after denitration reconstruction at a typical 300 MW "W" flame boiler, ash composition analysis results showed that except the Al_2O_3, SiO_2, Fe_2O_3 for conventional coal-fired fly ash components, it also contains a large amount of SO_3. This finding is closely related to the use of the high-sulfur coal, abnormal SCR denitration operation, and the ammonia consumption that has significantly exceeded design value.

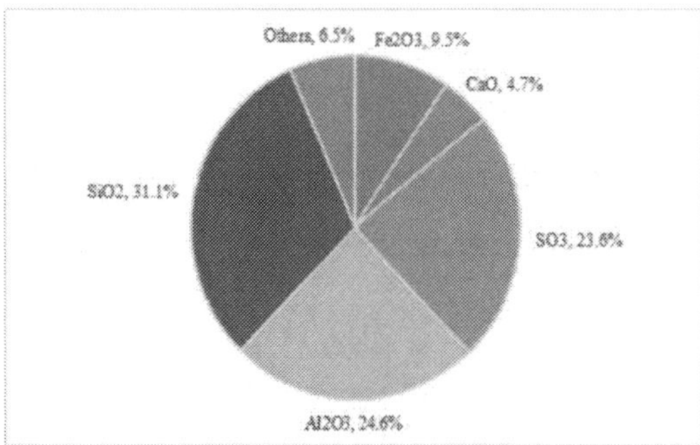

Fig.1. Composition analysis results of ash on the electrostatic precipitator at a specific 300MW unit.

Aimed at the problem of non-uniform flow field, boiler combustion optimization and tuning of denitration spraying ammonia test should be periodically carried out, guarantee the stability of the flow field in the SCR reactor. At the same time, it should also pay attention to the ammonia dampers, ammonia injection nozzle and other denitration equipment maintenance, so that the denitration device is efficient, stable and safe operation for a long time. When the inhomogeneity of the flow field is mild, the flow field in the denitrification reactor can be adjusted through field adjustment test to meet the operation requirements. As shown in Figure 2, the NOx concentration distribution at the reactor outlet is significantly improved after the optimal test of ammonia injection, which is more conducive to control the ammonia injection amount and ammonia escape concentration. However, when the flow field changes greatly or the design of denitrification device is not reasonable, the flow field problem cannot be eliminated simply through optimization test. In this case, it is necessary to carry out optimization and transformation of the flow field to solve the problem thoroughly.

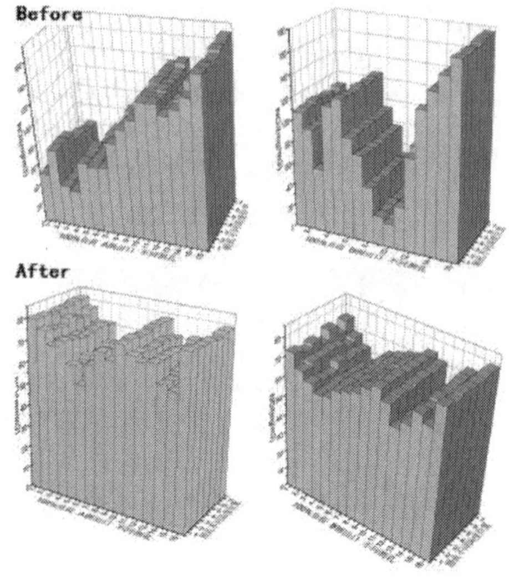

Fig.2. NOx concentration distribution before and after ammonia injection optimization test at a specific SCR denitrification facility.

3 Low-load Operation

At present, the contradiction between power supply and demand is increasingly serious in China. The unit load reduction and frequent peak adjustment happen frequently, which results in growing denitrification at low load operation. Before the transformation of ultra-low emission, the flue gas temperature at the outlet of the economizer under low-load conditions in some power plants was too low to meet the minimum continuous operation temperature of the catalyst, leading to the failure of the system in normal operation[3]. However, the minimum continuous operation temperature of the catalyst is determined by the concentrations of SO_3 and NH_3 in the flue gas. Due to the increase in the amount of ammonia injection required by ultra-low emission, the NH_3 concentration in the flue gas increases significantly, which further increases the minimum continuous operation temperature of the catalyst, and aggravates the problem of low-load denitrification operation. On the other hand, the environmental protection departments have not lowered their requirements for the above environmental protection indicators under abnormal working conditions, and some local governments have put forward explicit requirements for the depth peak regulation, wide load denitrification, standard discharge assessment, electricity price assessment and other aspects, thus posing great challenges to the operation stability of the denitrification system. As shown in Figure 3, based on the statistical results on 193 units NOx excessive emissions in 2017, the excess emissions caused by the insufficient of temperature at low load reached 11% and

14% respectively for number and time, which indicates abnormal denitration operation at low load has become an important factor affecting the stability of denitrification operation.

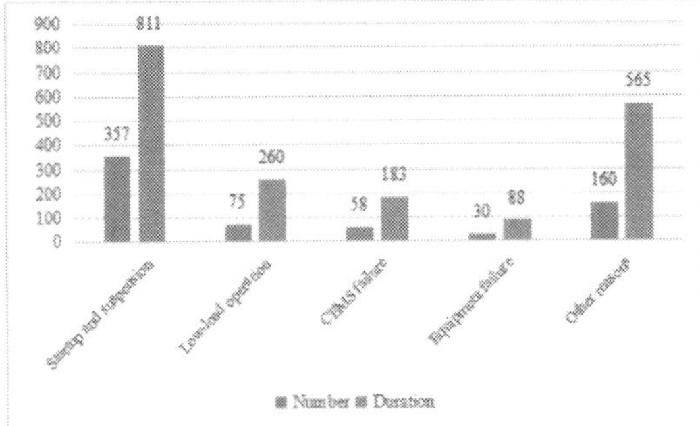

Fig.3. The reasons statistical result of NOx excessive emission

Based on the demand for wide load denitrification, some power generation enterprises have made active exploration in operation adjustment and achieved certain results[5]. The adjustment means mainly include boiler combustion adjustment, SCR operation mode adjustment and denitration operation control. Boiler combustion adjustment mainly covers the adjustment of the boiler air volume, the proportion of air distribution, burner Angle, coal mill operation level, coal mill outlet temperature, boiler long blowing, reheater side flue gas baffle degree and other means to improve flue gas temperature of the coal economizer outlet.The adjustment of SCR operation mode mainly include the efficient operation of the denitrification unit through the adjustment and optimization of ammonia injection. The formation concentration of NOx is minimized through the optimization of low-NOx combustion, so as to reduce the amount of ammonia injection and increase the minimum continuous operation temperature of the denitrification catalyst. Denitration operation control mainly lies in reducing the flue gas temperature limit of SCR exits on the basis of accurate verification on catalyst ABS (a lot of NH_4HSO_4 could be generated at this temperature), MIT (minimum ammonia injection temperature), MOT (minimum continuous operation temperature), and coordinating control unit load allocation and denitration operation on the basis of the analysis of the regularity of NH_4HSO_4 deposit and decomposition under specific flue gas conditions. In general, for projects with small flue gas temperature deviation from design value and large possibility to achieve low-load operation through optimized operation adjustment, factors affecting low-load operation of SCR should be comprehensively considered to form a feasible low-load operation mode to avoid large-scale transformation.

Current industry low-load denitration technology including economizer flue gas bypass, economizer feedwater bypass, economizer grading, hot water recirculation and so on, but the application of various technical measures all have certain boundary conditions, the investment and the impact on the running efficiency are different, so all items should be based on actual situation, comprehensive analysis of boundary conditions, in-depth analysis of the feasibility, applicability and economical efficiency of the retrofit scheme. The most optimal technical solutions should be selected after technical and economic comparisons. In particular, the retrofit project should adopt the most appropriate scheme according to the characteristics of the units in service, the condition of coal burning, the design parameters of SCR flue gas denitration system, the condition of equipment and the layout. In general, for units with small capacity and limited annual utilization hours, wide load denitrification technology with small amount of reforming works such as economizer flue gas bypass can be considered, while for units with large capacity and higher requirements for operation economy and reliability, priority should be given to technologies like economizer grading transformation.

4 On-line Instrument Accuracy

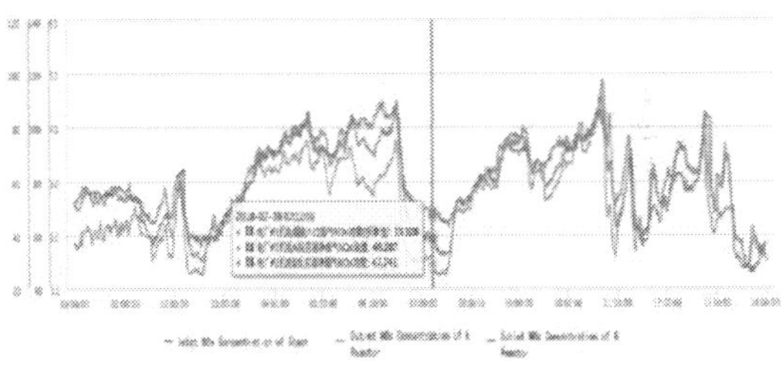

Fig.4. Inversion phenomenon of NOx monitoring result on a specific 600MW power unit

The NOx and NH_3 measuring instruments at the outlet of SCR denitrification reactor are the two most important instruments in the denitrification equipment. However, in the current practical operation, measurement results are often not representative due to limited instrument layout points, large variation range of boiler load and poor flow field[6]. As shown in Figure 4, the concentration of NOx at the SCR denitrification outlet is more than 2 times higher than that at the chimney inlet, which is a typical manifestation of unrepresentative NOx measurement points. On the other hand, the current SCR denitrification device requires that ammonia escape concentration below 3ppm. However, due to the harsh environment of flue gas, measurement technology limitations, and error transmission of sampling, analysis, calibration, data transmission and other aspects, the monitoring results of ammonia escape meter are generally inaccurate, which cannot play a guiding role in the operation of denitrification facilities[7]. As shown in Figure 5, 180 sets of ammonia escape meter running status show that 46 units of equipment have high failure rate, and another 8 units can not operate normally, that is to say, the abnormal operating unit accounted for 30%, and the other 70% that can run normally is subject to data accuracy verification.

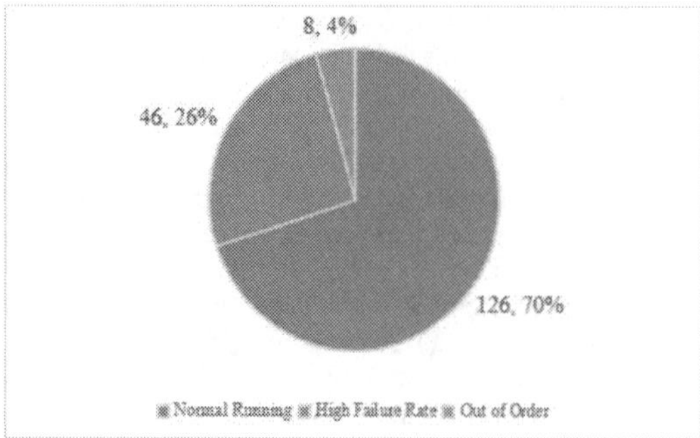

Fig.5. Running state statistical result of ammonia slip online monitoring device

In view of the lack of representativeness of the monitoring results of NOx meter at the outlet of the denitrification reactor, on the one hand, the actual operation should still be controlled according to the NOx concentration value at the chimney inlet considering that the NOx concentration field at the chimney inlet is generally uniform after "rectifying" by dust remover, desulfurization and other equipment. On the other hand, the assessment and optimization of the flow field in the denitrification reactor should be entrusted to the testing unit as soon as possible to ensure the uniformity of NOx concentration field at the reactor outlet. If necessary, the online instrument sampling points should be rearranged according to the grid test results to ensure the representativeness of the NOx online monitoring point. In addition, some power generation enterprises can effectively solve the problem of sampling representativeness by setting multi-point sampling devices and conducting mixed measurement.

In view of the above problems of the reliability of ammonia escape meter and the accuracy of monitoring results, power generation enterprises should strengthen the maintenance of monitoring instruments to ensure that the monitoring instruments can reflect the variation trend of ammonia escape. At the same time, the ammonia escape situation can be comprehensively determined by overall consideration of the online parameter changes of outlet NOx concentration, ammonia consumption and air preheater pressure difference. If conditions permit, the regular monitoring of ammonia content in fly ash by ion chromatography (DL/T 1494-2016) can be referred to judge the change of ammonia escape.

5 Catalyst Life

The denitration efficiency requirements are higher than 85% generally after ultra-low emission reconstruction, requirements on flow field uniformity and catalyst performance stability further improve. The catalyst quality problem, fly ash block and wear, flow field inhomogeneity bring about a large number of catalyst replacement needs, which generally fail to meet the designed chemical life[8]. In addition, although the ultra-low emission standard requires the nitrogen oxide emission concentration to be controlled below $50mg/m^3$, in actual operation, most power plants generally control the emission at about $30mg/m^3$ to avoid system fluctuation and exceed the standard, further shortening the service life of the catalyst. As shown in Figure 6, the statistical results of seven ultra-low emission units with the need to replace catalysts show that the actual service life of all units is shorter than the chemical life guaranteed by performance, and the shortest actual service life is only 44% of the guaranteed performance life.

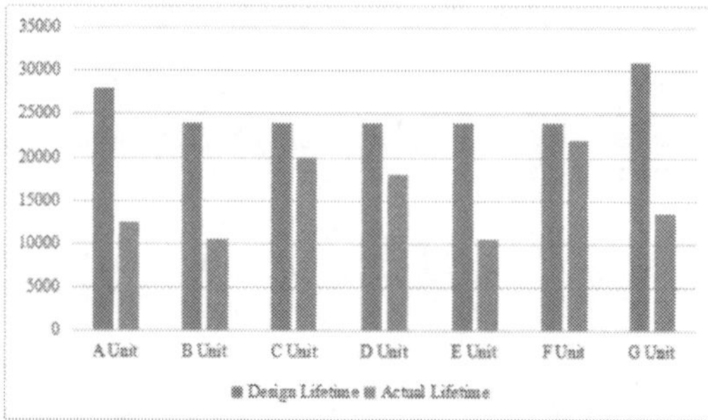

Fig.6. Service life statistical result of the proposed replacement catalyst

In denitration project design and selection design of catalyst, catalyst performance and life expectancy is designed according to the boundary conditions and the performance guarantee values. In the actual operation, the service life of catalyst is affected by the factors such as the coal and the unit operation condition which often deviates from the design conditions. Thus, the catalyst life management work need to be adjusted according to actual situations[9]. But this work is demanding in professional equipment and personnel skills given a confluence of factors. For example, the practical denitration catalyst operation seldom requires adjustment and control of the power generation enterprise operations staff. The catalyst control is quite difficult to implement due to the concealment of the catalyst life problem. Besides, there are inaccurate measurement of ammonia escape, non-uniform flow field and other related problems. Therefore, in the process of SCR denitration operation, professional organizations should be invited to help solve the above problems, technical services of catalyst life management should be carried out.

Service contents of catalyst life management technology mainly include carrying out evaluation test, diagnosis test, optimization test, in-service catalyst performance test, catalyst life evaluation and making prediction on a regular or irregular basis. Evaluation test is implemented to know the performance status of denitrification device and identify problems in time. When there is a problem, the diagnostic test is launched to figure out the causes. To solve the problems of ammonia escape exceeding the standard and uneven flow field operation, optimization test should be carried out. The plan of catalyst addition, replacement, regeneration or disposal should be formulated based on the attenuation of catalyst activity resulting from regular catalytic performance testing and the predicted residual life of catalysts,obtained from the evaluation results. The denitration device operation problem can be solved through catalyst life management technology service. Corresponding prevention and solution measures are put forward to eliminate problems at an early stage, thus reducing denitration device running cost and prolonging the life of denitration catalyst so as to ensure that the denitration device to maintain safe, stable and efficient operation and reduce the environmental risk.

6 Conclusion

In view of the current stability and reliability problems in the ultra-low emission operation of NOx in coal-fired power plants, power generation enterprises, relevant industrial companies and technical service units should timely summarize the causes and lessons of the stability and reliability problems of denitrification devices, and form corresponding countermeasures. Power generation enterprises should strengthen the technical training of operational management, improve operations staff so as to prevent, judgment, analysis and the ability to solve the problem of denitration operation. Technical service units should provide quality and efficient technical service, discover and solve the problems timely in the operation of the denitration device and ensure the efficient operation of the denitration device. Related industry companies should provide quality products and economic and effective engineering optimization. By giving full play to the expertise of all parties, the stable, efficient and reliable operation of the NOx ultra-low emission in coal-fired power plants can be realized.

We appreciate the financial support from the National Key Research and Development Program (NO. 2017YFB0603201) and the Key Science and Technology Projects of China Huadian Group Co., Ltd. (CHDKJ17-01-55).

References

1. ZHU Yue. Key technical issues of environmental protection island of thermal power plant [J]. Power Generation Technology, 2018, 39(1): 1-12.
2. ZHANG Yang, FENG Qianwei, PEI Yukun, et al. Study on operation&maintenance technology of SCR denitrification from coal-fired boiler flue gas [J]. Electric Power Technology and Environmental Protection, 2016, 32(5): 32-34.
3. ZHANG Yang, YANG Yonglong, FENG Qianwei, et al. Key technical issues of SCR denitrification from coal-fired boiler flue gas [J]. Electric Power, 2015, 48(4): 32-35.
4. LI Tong. Ultra-low emission reconstruct of denitrification system based on NOx uniformity test of SCR reactor [J]. Chinese Journal of Environmental Engineering, 2018, 12(4): 1112-1119.
5. SUN Weipeng, CHEN Shaohua, JIANG Yong, et al. Full load operation practice of a denitration unit with zero retrofit on the equipment [J]. Power Equipment, 2018, 32(3): 184-187.
6. LI Qing, JIANG Long, GUO Yue, et al. Current applications and key problems of ultra-low emission for coal-fired power plant [J]. High Voltage Engineering, 2017, 43(8): 2631-2637.
7. LI Chao, ZHANG Yang, ZHU Yue. Analysis on the application status of ammonia slip online monitoring of SCR denitrification from coal-fired boiler flue gas [J]. Power Generation & Air Condition, 2017, 38(5): 41-44.
8. DU Zhen, WANG Zhidong, YAN Min, et al. Analysis on actual operation performance of SCR denitration catalyst [J]. Electric Power, 2018, 51(2): 133-136.
9. ZHOU Jianxin, YU Cong, JIANG Xiaoming, et al. System for performance evaluation and lifetime management of SCR DeNOx catalysts in coal-fired power plants [J]. Electric Power, 2015, 48(4): 11-15.

E3S Web of Conferences **120**, 03003 (2019) https://doi.org/10.1051/e3sconf/201912003003
CGEEE 2019

Cu adsorption in fixed bed column with three different influent concentration

Huang-Mu Lo[1,*], *Kae-Long* Lin[2], *Min-Hsin* Liu[1], *Hsung-Ying* Chiu[1], and *Fang-Cheng* Lo[3]

[1]Department of Environmental Engineering and Management, Chaoyang University of Technology, 41349 Taichung, Taiwan.
[2]Department of Environmental Engineering, Ilan University, 26047 Ilan, Taiwan.
[3]Graduate Institute of Environmental Engineering, National Taiwan University, 10617 Taipei, Taiwan.

Abstract. Heavy metals from the electroplating wastewater might cause environmental pollution if not well treated. Generally, carbon adsorption might be used for the final step for further trace metals removal. This study investigated the heavy metal Cu adsorption in the fixed bed column with 1, 10 and 100 mg/L influent concentration. Results showed that K_{AB} decreased as influent Cu concentration increased from 1 to 100 mg/L while N_0 increased as influent concentration increased from 1 to 100 mg/L as can be found in Adams-Bohart model. R^2 was found between 0.8579 and 0.9182. In Thomas model. K_{TH} and q_0 showed the similar trend as K_{AB} and N_0 in the Adams-Bohart model. K_{TH} decreased as influent Cu concentration increased from 1 to 100 mg/L. q_0 increased as influent Cu concentration increased from 1 to 100 mg/L. R^2 of regression model was found between 0.9065 and 0.9836. In Yoon-Nelson model. K_{YN} increased as influent Cu concentration increased from 1 to 100 mg/L while τ decreased as influent Cu concentration increased from 1 to 100 mg/L. Results showed that the three models of Adams-Bohart model, Thmoas model and The Yoon-Nelson model were suitable for the description of Cu adsorption by activated carbon.

1 Introduction

Heavy metals such as Cd, Cr, Cu, Ni, Pb and Zn in the wastewater effluent might cause the environmental and health risks if not appropriately treated [1-3]. Wastewater might be treated with physical, chemical and biological processes [4-6] to remove the pollutants such as COD, SS, and heavy metals. Generally, the process for wastewater treatment includes primary sedimentation, active sludge process, secondary sedimentation, coagulation, flocculation, filtration, adsorption and chlorination etc. Among them, adsorption [4] is mostly used as final process to remove the trace organic matters and heavy metals. Organic pollutants and metals adsorption were investigated by batch and fixed bed column to understand the removal of contaminants. Biosorbent was used to examine the adsorption capacity with continuous fixed bed column [7]. Akhigbe et al. (2016) investigated the disinfection and removal performance for Escherichia coli and heavy metals by silver-modified zeolite in a fixed bed column [8]. Coalesced chitosan activated carbon composite for batch and fixed-bed adsorption of cationic and anionic dyes was also represented by Auta and Hameed [9]. Evaluation of the adsorptive capacity by peanut hull pellets for heavy metals in solution was also reported [10]. Adsorption and detection of organic pollutants by fixed bed carbon using nanotube electrochemical membrane was examined by Buffa and Mandler [11]. A review of adsorption of heavy metals on conventional and nanostructured materials for wastewater treatment purposes was intensively reported by Burakov et al. [4]. Biosorption of Zn(II) from industrial effluents using sugar beet pulp and F. vesiculosus with laboratory and pilot approach was studied by Castro et al. [12]. Chao et al. [13] reported the biosorption of heavy metals on citrus maxima peel, passion fruit shell, and sugarcane bagasse for copper(II), cadmium(II), nickel(II), and lead(II) metal ion removal in a fixed-bed column. Chatterjee et al. [14] reported adsorptive removal of potentially toxic metals (cadmium, copper, nickel and zinc) by chemically treated laterite: Single and multicomponent batch and column study. Ding et al. [15] investigated the removal of lead, copper, cadmium, zinc, and nickel from aqueous solutions by alkali-modified biochar: Batch and column tests. Ghasemabadi et al. [16] presented the continuous adsorption of Pb(II), As(III), Cd(II), and Cr(VI) using a mixture of magnetic graphite oxide and sand as a medium in a fixed-bed column. Kavand et al. [17] reported the adsorption of heavy metal ions including lead (Pb^{2+}), Cadmium (Cd^{2+}) and Nickel (Ni^{2+}) onto a commercial activated carbon (AC) in single and multi-component aqueous fixed bed column. Different adsorbents for different metals and organic pollutants were investigated as mentioned in the above literature.

This study aims to investigate the breakthrough curve of Cu adsorption by activated carbon via three different influent concentrations using three models (Adams-

* Corresponding author: hmlo@cyut.edu.tw

Bohart model, Thmoas model and The Yoon-Nelson model).

2 Materials and methods

The working fixed bed column was 25 cm high and the diameter was 2 cm wide. Thus, the bed volume (BV) was 78.54 cm^3. Activated carbon was used as adsorbent. Three influent Cu concentrations of 1, 10, and 100 mg/L were used as adsorbate. Influent velocity was 25 cm/min. The adsorption experiment was conducted in room temperature about 25 ℃. Characteristics of activated carbon can be seen in Table 1.

The filtrate was collected at each BV such as 1, 2, 3 etc. till the effluent Cu concentrations (C_t) was equal to the influent Cu concentrations (C_0). Plot the BV against the C_t/C_o ratio, the breakthrough curve can be obtained.

Three dynamic adsorption models were used to describe the adsorption behaviour of Cu adsorption on activate carbon. Three dynamic adsorption models (8, 16) were expressed as following three equations:

Adams-Bohart model:

$$\ln(C_t/C_0) = K_{AB}C_0t - K_{AB}N_0(Z/F) \quad (1)$$

where K_{AB} = adsorbed dynamic constant (L/mg-min)
C_t/C_0 = Effluent concentration/Influent concentration
N_0 = Saturated concentration (mg/L)
Z = Height of fixed bed column (cm)
F = Flow velocity of adsorbate (cm/min)

Thomas model:

$$\ln((C_0/C_t)-1) = K_{TH}q_0W/Q - K_{TH}C_0t \quad (2)$$

where K_{TH} = adsorbed dynamic constant (L/mg-min)
C_0/C_t = Influent concentration/Effluent concentration
W = weight of adsorbent (g)
q_0 = adsorbate adsorbed by adsorbent (mg/g)
Q = Influent flow rate (L/min)

$$\ln(C_t/(C_0 - C_t)) = K_{YN}t - \tau K_{YN} \quad (3)$$

where K_{YN} = adsorbed dynamic constant (1/min)
τ = time as 50% adsorbate Cu adsorbed on activated carbon (t)

Table 1. Characteristics of activated carbon.

Parameters	
Particle size	3~4 mm^3
Ball-pan Hardness	≧ 95%
Moisture as packed	≧ 8%
Ash	≧ 5%
Iodine Number	≧ 1050 mg/g
Total Surface Area(BET, N_2)	≧ 1100 m^2/g
Bulk density	0.4~0.55 g/cc
pH of Water Extract	7-11

Using equation (1) ~ (3), plot the Y axis against the X axis (BV), the regression lines were obtained. The parameters in the equations can be obtained as shown in Table 2-4.

3 Results and discussion

Parameters of three models can be found in Table 2-4.

Table 2. Adams–Bohart model-Cu.

Influent conc. (mg/L)	Bed height (mm)	Velocity (mL/min)	K_{AB} (L/mg-min)	N_0 (mg/L)	R^2
1	250	25	0.0296	76.85	0.8579
10	250	25	0.0070	292.84	0.8744
100	250	25	0.0003	2715.92	0.9182

Table 2 showed the parameters of Adams-Bohart model. K_{AB} decreased from 0.0296 to 0.003 L/mg-min as influent concentration increased from 1 to 100 mg/L while N_0 increased from 76.85 to 2715.92 mg/L as influent Cu concentration increased from 1 to 100 mg/L. R^2 was found between 0.8579 and 0.9182.

Table 3. Thomas model-Cu.

Influent conc. (mg/L)	Bed height (mm)	Velocity (mL/min)	K_{TH} (L/min-mg)	q_0 (mg/g)	R^2
1	250	25	0.0442	92.46	0.9065
10	250	25	0.0190	271.34	0.9836
100	250	25	0.0019	1432.41	0.9361

Table 3 showed the parameters of Thomas model. K_{TH} and q_0 showed the similar trend as K_{AB} and N_0 in the Adams-Bohart model. K_{TH} decreased from 0.0442 to 0.0019 L/min-mg as influent Cu concentration increased from 1 to 100 mg/L. q_0 increased from 92.46 to 1432.41 mg/g as influent Cu concentration increased from 1 to 100 mg/L. R^2 of regression model was found between 0.9065 and 0.9836.

Table 4. The Yoon–Nelson model-Cu.

Influent conc. (mg/L)	Bed height (mm)	Velocity (mL/min)	K_{YN} (1/min)	τ (min)	R^2
1	250	25	0.043	53.23	0.9065
10	250	25	0.1898	15.25	0.9836
100	250	25	0.1889	8.05	0.9361

Table 4 showed the parameters of the Yoon-Nelson model. K_{YN} increased as influent Cu concentration increased while τ decreased as influent Cu concentration increased. K_{YN} increased from 0.043 to 0.1889 1/min as influent Cu concentration increased from 1 to 100 mg/L while τ decreased from 53.23 to 8.05 min as influent Cu concentration increased from 1 to 100 mg/L. The τ (53.23 min) needed to be half saturated in the influent Cu concentration of 1 mg/L was ~3.5 times that of 10 mg/L influent Cu concentration and ~7 times that of 100 mg/L influent Cu concentration.

The breakthrough curves of 1, 10 and 100 mg/L Cu influent concentration and the regression of the Adams-Bohart model, Thomas model and The Yoon–Nelson model was shown in Figure 1-3.

Figure 1 showed the influent Cu 1 mg/L adsorption breakthrough curve and the three dynamic adsorption model and regression line. Results was presented in Table 2-4.

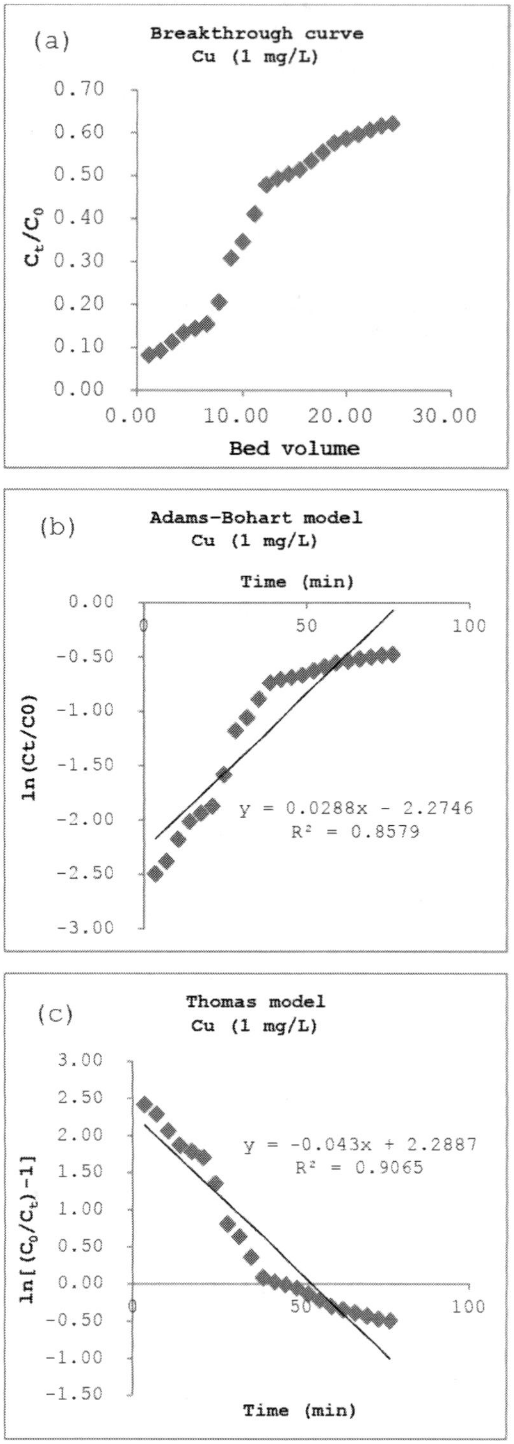

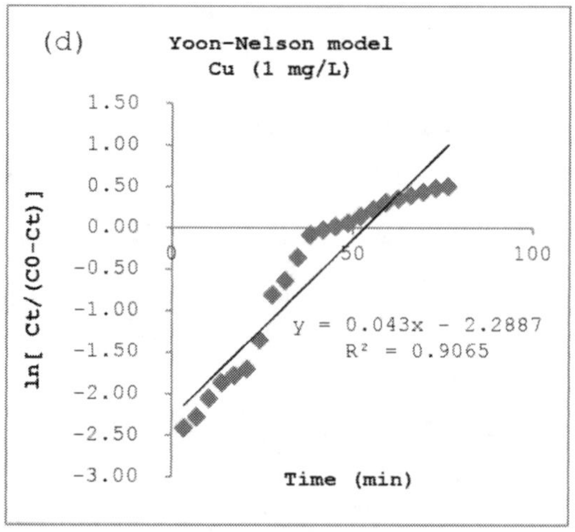

Fig. 1. Cu adsorption breakthrough curve and three dynamic adsorption model in fixed bed column (Cu: ~1 mg/L, column height: 25 cm, flowing velocity: 25 cm/min)

Figure 2 showed the influent Cu 10 mg/L adsorption breakthrough curve and the three dynamic adsorption model and regression line. Results was presented in Table 2-4.

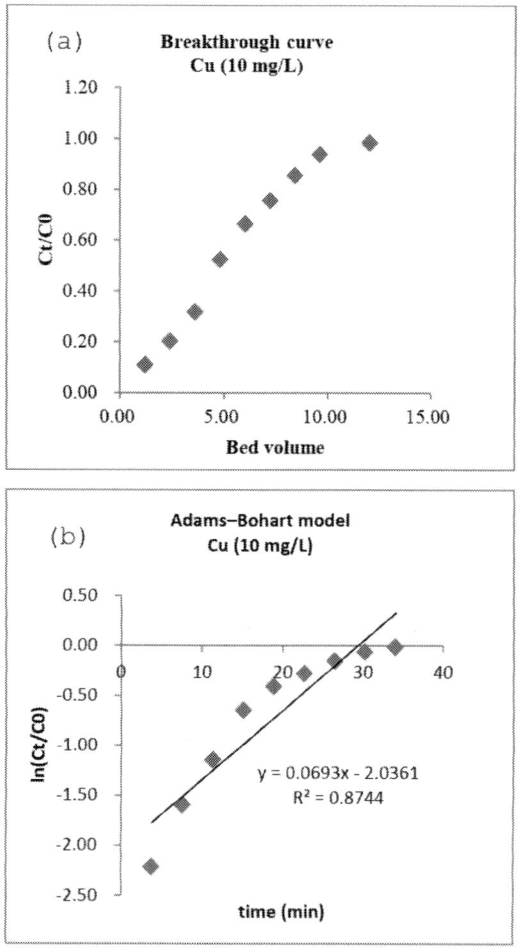

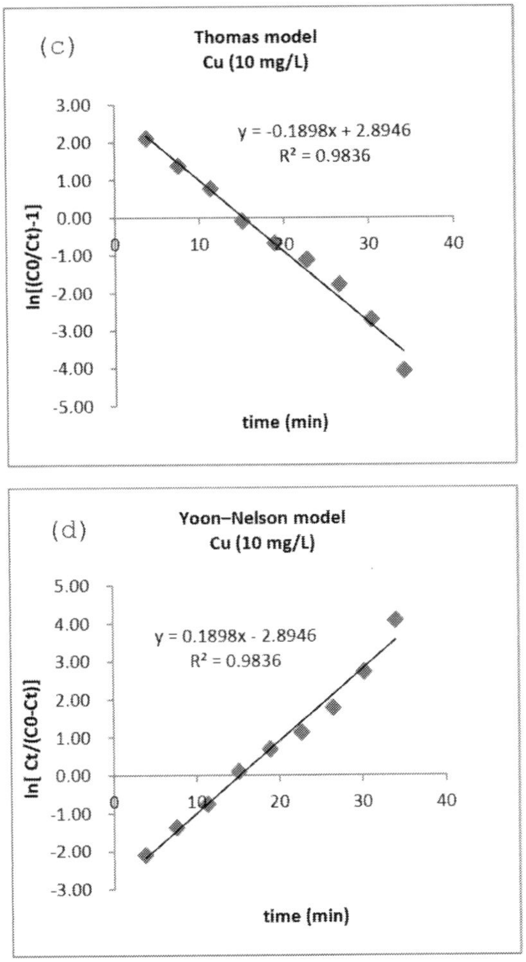

Fig. 2. Cu adsorption breakthrough curve and three dynamic adsorption model in fixed bed column (Cu: ~10 mg/L, column height: 25 cm, flowing velocity: 25 cm/min)

Figure 3 showed the influent Cu 100 mg/L adsorption breakthrough curve and the three dynamic adsorption model and regression line. Results was presented in Table 2-4.

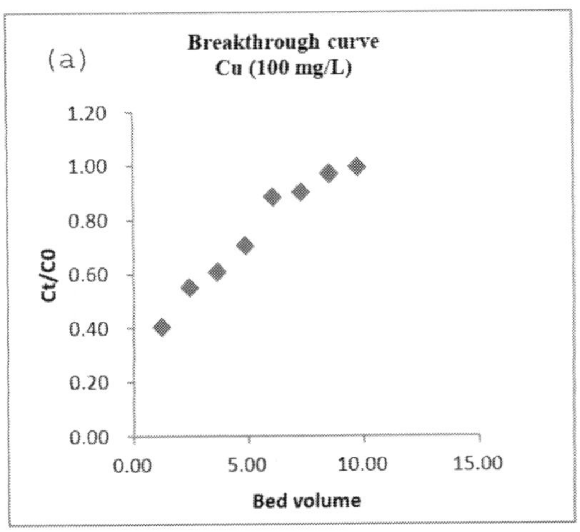

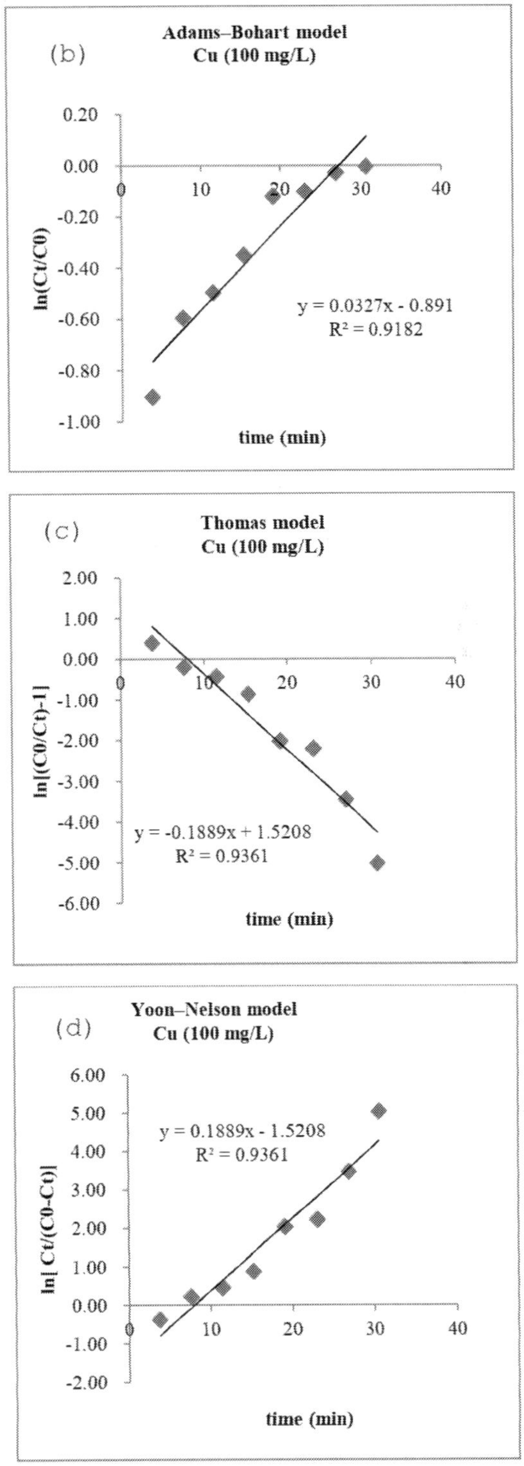

Fig. 3. Cu adsorption breakthrough curve and three dynamic adsorption model in fixed bed column (Cu: ~100 mg/L, column height: 25 cm, flowing velocity: 25 cm/min)

Results showed that Cu adsorption in fixed bed column were described well in the Adams-Bohart model, Thmoas model and The Yoon-Nelson model.

4 Conclusions

Activated carbon was used to adsorb the influent Cu concentration of 1, 10 and 100 mg/L in fixed bed column. Results showed that breakthrough curve was found quick to reach saturation point in influent Cu 100 mg/L concentration than in influent Cu 1 and 10 mg/L concentration. Results also showed that the three models of Adams-Bohart model, Thmoas model and The Yoon-Nelson model were suitable for the description of Cu adsorption.

References

1. B. Hu, S. Shao, Z. Fu, Y. Li, H. Ni, S. Chen, Y. Zhou, B. Jin, Z. Shi, Sci. Total Environ. 658, 614-625 (2019)
2. Q. Yang, Z. Li, X. Lu, Q. Duan, L. Huang, and J. Bi, Sci. Total Environ. 642, 690-700 (2018)
3. L. Zhang, G. Zhu, X. Ge, G. Xu, and Y. Guan, J. Hazard. Mater. **360**, 32-42 (2018)
4. A.E. Burakov, E.V. Galunin, I.V. Burakova, A.E. Kucherova, S. Agarwal, A.G. Tkachev, V.K. Gupta, Ecotoxicol. Environ. Saf. **148**, 702-712 (2018)
5. D. Wang, G. Tang, Z. Yang, X. Li, G. Chai, T. Liu, X. Cao, B. Pan, J. Li, G. Sheng, X. Zheng, Z. Ren, J. Hazard. Mater. (2019), https://doi.org/10.1016/j.jhazmat.2019.03.069
6. L. Wang, Y. Wang, F. Ma, V. Tankpa, S. Bai, X. Guo, X. Wang, Sci. Total Environ. **668**, 1298-1309 (2019)
7. A. Abdolali, H.H. Ngo, W. Guo, J.I. Zhou, J. Zhang, S. Liang, S.W. Chang, D.D. Nguyen, Y. Liu, Bioresour. Technol. **229**, 78-87 (2017)
8. L. Akhigbe, S. Ouki, and D. Saroj, Chem. Eng. J. **295**, 92-98, (2016)
9. M. Auta and B. H. Hameed, Colloids Surf B Biointerfaces **105**, 199-206, (2013)
10. P. Brown, I. Atly Jefcoat, D. Parrish, S. Gill, and E. Graham, Adv. Environ. Res. **4**, 19-29 (2000)
11. A. Buffa and D. Mandler, Chem. Eng. J. **359**, 130-137 (2019)
12. L. Castro, M. L. Blázquez, F. González, J. A. Muñoz, and A. Ballester, Sci. Total Environ. **598**, 856-866 (2017)
13. H.-P. Chao, C.-C. Chang, and A. Nieva, J. Ind. Eng. Chem. . **20**, 3408-3414, (2014)
14. S. Chatterjee, I. Sivareddy, and S. De, J. Environ. Chem. Eng. 5, 3273-3289 (2017)
15. Z. Ding, X. Hu, Y. Wan, S. Wang, and B. Gao, J. Ind. Eng. Chem. **33**, 239-245 (2016)
16. S. M. Ghasemabadi, M. Baghdadi, E. Safari, and F. Ghazban, J. Environ. Chem. Eng. **6**, 4840-4849 (2018)
17. M. Kavand, E. Fakoor, S. Mahzoon, and M. Soleimani, Process Saf. Environ. Prot. **113**, 330-342 (2018)

Machine learning algorithms for predicting air pollutants

Jirat Boonphun, *Chalat* Kaisornsawad, and *Papis* Wongchaisuwat*

Industrial Engineering Department, Kasetsart University, 50 Ngamwongwan Rd, Ladyao Chatuchak Bangkok, Thailand

Abstract. An atmospheric particular matter, commonly recognized as PM, contains solid particles and liquid droplets suspending in an ambient air. A high concentration of PM is known to seriously cause adverse health effects to humans especially a small-sized particle, known as PM2.5. Not only health effects, environmental effects are also obviously observed. This work aims to estimate a likelihood of PM2.5 exceeding a pre-defined safety threshold. Multiple machine learning models are explored in this work. Particularly, classification models are implemented based on meteorological data and air pollutant features measured at different altitudes above a ground level. These features are shifted back to various time steps resulting in more insightful time-lagged features. Furthermore, a feature selection technique is implemented to specify a desirable set of important features. A re-sampling technique is also employed to address an unbalancing level of the response value in an original data set. The proposed models are evaluated on a case study whose data set is collected from an air monitoring station located in Bangkok, Thailand.

1 Introduction

A particular matter or PM represents a mixture of solid and liquid such as dust and ash in the air. It normally varies in a diameter or a width of the particle. In this study, we focus on PM2.5 referring to an atmospheric particular matter whose size is generally less than 2.5 micrometres. Fine particles like PM2.5 are more likely to stay longer in the air compared to other heavy particles. As a result, humans have a higher chance to inhale these fine particles into their bodies leading to adverse health effects [7]. A strong relationship between an exposure to fine particles and a mortality risk from lung or heart diseases [18] is observed. Once an amount of PM2.5 reaches a healthy threshold, it is strongly advisable to take an action to protect human health.

Most crowded cities are prone to face air pollution problems. There is no exception for Bangkok, the capital city of Thailand. According to the study by Chuersuwan et. al. [5], primary sources of PM2.5 in Bangkok are generally an automobile and a biomass burning in both traffic and residential sites. Primary problems include smog and respiratory issues whose severity has become worse during the past few years. Recently, Bangkok experienced so much smog due to PM2.5 that the government had to issue a serious health warning to residents. These motivate us to develop models to estimate an hourly concentration of an air pollutant in advanced. As a safety precaution, these predictive models are considered as a basis for applying successful pollution control processes.

Significant factors potentially impacting air pollutant concentrations include meteorological conditions such as an air temperature, a wind speed, a wind direction as well as a relative humidity [8]. High concentrations of traffic-related pollutants are strongly correlated to a low wind speed. Moreover, high concentrations of the particular matter are commonly related to high humidity in the air [10]. We further speculate that a relative humidity has a strong relationship with an amount of precipitation which in turn impacts air pollutant concentrations. In addition, we are interested in studying how PM2.5 changes with respect to other air pollutants. Due to these reasons, meteorological data and various air pollutants are utilized as initial features in forecasting PM2.5 concentrations.

In this study, we pay more attention to a situation where the amount of PM2.5 exceeds a pre-defined healthy threshold. In response, classification models are proposed in order to predict whether PM2.5 is above the safety limit. These models rely on initial features including meteorological data and multiple air pollutants which are further utilized to generate additional features. In particular, these features are shifted to different time periods resulting in various time-lagged features. Due to time-lagged features, significant number of features are retrieved as a result. In order to address this issue, a feature selection technique is implemented which controls total number of features used in the models.

We also observe that a number of instances where PM2.5 value exceeds the threshold is relatively low. In response, we employ a re-sampling technique to up-sample the minority class and down-sample the majority class prior to actually fitting the models. Classifiers including Naïve Bays, Logistic regression, Random forest, and Neural networks are implemented. To evaluate the performance of the models, traditional metrics such as a confusion matrix and F1-score are utilized.

As a case study in this work, a data set collected from an air monitoring station at Kasetsart University located in Bangkok is used to train and test the models. While

* Corresponding author: fengppwo@ku.ac.th

PM2.5 is measured at a fixed height above a ground level, meteorological and air pollutant features are collected at various altitudes. Instead of focusing only at ground-level information, the data set is specifically gathered along a vertical basis, i.e. at 30, 75, and 110 meters above the ground. We speculate that data collected at different heights potentially contains useful insights in forecasting the PM concentrations. Hence, we consider information measured at 30 meters as a low level while the rest are counted as a high level. To verify our assumption, features with high score due to the feature selection technique are thoroughly observed.

Features as well as an application of the models distinguish our work from others. Our main contribution is exploring other insightful features that have not been studied in prior works. Particularly, we consider meteorological and air pollutant concentrations locally measured at different heights above the ground. In addition, we introduce features at various time steps which potentially capture deeper insights. The feature selection technique is also performed to include only significant features. To address an imbalance problem observed in our data set, the re-sampling technique is employed. To the best of our knowledge, none of prior works applied similar machine learning algorithms based on a data set collected in Bangkok. Our data set is specifically gathered at the particular air monitoring station in an exceptional fashion, i.e., measuring information at 3 different heights above the ground.

We summarize the literature in Section 2. The features, the models and evaluation metrics used to test the models are thoroughly discussed in Section 3. Further descriptions regarding data collections and preparations are illustrated in Section 4. Results of our algorithms including all experiments are reported in Section 5. Discussions are provided in Section 6 while a conclusion is stated in the last section.

2 Related work

Air pollution is one of the most critical issues in many cities globally due to its adverse health effects. Bellinger et. al. [2] conducted a systemic review of applying data mining and machine learning techniques in an air pollution epidemiology. According to [2], all 400 reviewed articles are separated into 3 main research areas which include a source apportionment, a prediction of air pollution or quality or exposure, and a hypothesis generation. Our work closely aligns with the second category. An overview of various air pollution forecasting algorithms is provided in [1]. Specifically, Bai et. al. extensively reviewed theory and applications of multiple predictive models as well as further compared advantages and disadvantages among models.

The particular matter or PM especially PM2.5 is commonly counted as one of the most detrimental pollutants posing a great threat to human health. Therefore, our work primarily focuses on PM2.5 only. Substantial research has been done for developing predictive models to estimate the PM2.5 concentrations in advanced. Most of these work aim to directly forecast the value of the PM2.5 [17] [16] [9] [19] [6] [14] [12] [11] [21] [20] [15]. Instead, our study focuses on determining a likelihood of the PM2.5 exceeding the safety limit. Another area of research that is closely related to us is predicting the ground PM2.5 concentrations based on satellite data. Significant number of works had been done along this line as summarized in [4]. Comparing with these works, our study is based only on locally collected features.

A few research have been done for speculating air pollutant concentrations in Thailand. Kanabkaew [13] predicted hourly PM concentrations in Chiangmai, Thailand based on MODIS and ground-based meteorological data. While temperature and relative humidity as ground-based meteorological data and satellite data are used in [13], our work relies on additional ground-based information and other pollutant concentrations. Furthermore, [13] performed simple and multiple linear regression models but we instead use more complex classification models for predicting the likelihood of PM2.5 exceeding the threshold. In addition, Amphanthong and Busbabodhin conducted a study to predict PM10 in the northern part of Thailand based on Grey system [22].

3 Methodology

In this study, we aim to estimate an hourly fine particular matter (PM2.5) concentration in Bangkok, Thailand. Most urban populations can be adversely affected by a greatly exposure to air pollutants like PM2.5 in an ambient air. We pay more attention to the case where the pollutant concentrations are above an air quality standard. Our problem is scoped down to identify whether the PM2.5 concentration exceeds a given threshold. Several classification models are implemented.

Our proposed models are based on meteorological data measured at different height above a ground surface as well as other air pollutants as initial features. Taking into account different time period of these features results in large number of features. We hence apply a dimensionality reduction technique before applying the classification models. The re-sampling technique is also implemented to handle an imbalance data set like ours. Evaluation metrics are discussed in detailed next.

3.1 Features

Meteorological and air pollutants features used in the models are initially collected at 3 different heights which are 30, 75, and 110 meters above a ground level. At each level, we gather 4 meteorological features including Wind Speed (WS), Wind Direction (WD), Temperature (TEMP), and Relative Humidity (RH). In addition, 6 air pollutants consisting of Nitric Oxide (NO), Nitrogen Dioxide (NO_2), Nitrogen Oxide (NOX), Sulfur Dioxide (SO_2), Carbon Monoxide (CO), and Ozone (O_3) are considered. In summary, there are 10 features for each height level resulting in 30 features in total.

We consider multiple time steps corresponding to these features. In particular, a hyperparameter n_step is defined as a number of time steps which we shift each feature back. According to a statistical test, F-value is used to measure a level of dependency between features and the response. We introduce a hyperparameter n_feature to specify a desirable number of features. In particular, we construct the final set of features by selecting top n_feature with highest F-values.

3.2 Re-sampling

The likelihood of PM2.5 exceeding the safety threshold is low among all instances in the data. Due to an imbalance of the data set, it is relatively challenging for any algorithm to provide a good performance. We employ a re-sampling technique to reduce the unbalancing level within the original data set in order to enhance the predictive power of the models. Specifically, Synthetic Minority Over-sampling Technique (SMOTE) is used to oversample the minority class and undersample the majority class [3]. For over-sampling, a random point along a line segment between the minority class sample and its k nearest neighbors is constructed. On the other hand, the majority class samples are randomly removed in order to under-sample. In our case, PM2.5 exceeding the pre-defined threshold is considered as the minority class while the rest is the majority.

3.3 Classification models and evaluation metrics

We utilize classification models to predict the likelihood of PM2.5 exceeding the safety threshold. Multiple supervised learning algorithms including Naïve Bays, Logistic regression, Random forest, and Neural networks are experimented with. The predicted probability associated with each class is achieved from the classification models. Multiple cut-off probabilities to identify the predicted class are also thoroughly investigated. We then compute traditional metrics such as the confusion matrix as well as the F1-score to evaluate the performance of these models. In order to compare among models, the F1-score is mainly considered. Different hyperparameters are calibrated to maximize the performance of the models.

4 Data preparation

All meteorological data and air pollutants are collected from the KU tower located at Kasetsart University, Bangkok, Thailand. Data collectors are installed at 30, 75 and 110 meters above the ground level in order to measure meteorological information and air pollutants in a vertical basis. These raw data are collected at every minute from 2015 to early 2019. We use an average value of each minute in the same hour as a representation of an hourly data. In summary, our data set has 36624 instances in total with 30 features at each time stamp. Among these instances, there are 2203 cases whose PM2.5 concentrations are above the threshold value which is set to 50 micrograms per cubic meter ($\mu g/m^3$). We label this case as "1" class in our binary classification model. This is accounted for approximately 6 percent.

In the original data set, numerous values are missing or irregular due to an error from data collectors. In practice, irregularities can be handled differently such as replacing with 0 or any statistical value. In contrast, our data set is in a chronical order; therefore, we propose to replace each irregularity with some default value. These default values are specifically computed based on an average of all remaining values in the last 3 hours after ignoring irregularities.

5 Results

We calibrate the hyperparmeters in order to obtain the best models' performance. A ratio between the number of samples in the majority class over the number of samples in the minority class after resampling with respect to the SMOTE technique is set to 0.5. For the neural network classification model, 1 hidden layer with 512 hidden nodes are selected after extensively experiments. The cut-off probability of 0.5 is particularly chosen.

F-value is used to measure the significance of each feature in the models. Table 1 illustrates top 30 features based on F-value. Top n_feature with high F-value are finally selected. Different values of n_step and n_feature are experimented among all 4 models while F1-score is used as an evaluation metric. A comparison of F1-score among different models and their corresponding parameters is shown in Table 2. Additionally, the confusion matrix evaluated on 1832 instances in a test set corresponding to the best model, i.e. the Random forest with 1 time step employing top 20 features is provided in Table 3.

Table 1. Top 30 features with high F-value

PM2.5(t-1)	H30m_ NO2(t-3)	H30m_ NOX(t)	H110m_ SO2(t-3)	H75m_ SO2(t-1)
PM2.5(t-2)	H75m_ NO2(t-1)	H75m_ NO2(t-2)	H110m_ NO2(t-1)	H110m_ NO2(t-2)
PM2.5(t-3)	H110m_ SO2(t)	H110m_ SO2(t-2)	H30m_ NOX(t-3)	H75m_ NOX(t-2)
H30m_ NO2(t-1)	H75m_ NO2(t)	H30m_ NOX(t-2)	H75m_ SO2(t)	H75m_ SO2(t-2)
H30m_ NO2(t)	H110m_ SO2(t-1)	H75m_ NO2(t-3)	H75m_ NOX(t-1)	H75m_ SO2(t-3)
H30m_ NO2(t-2)	H30m_ NOX(t-1)	H110m_ NO2(t)	H75m_ NOX(t)	H110m_ NO2(t-3)

Table 2. A comparison of F1-score among models and their corresponding hyperparameters

Model	n_step	n_feature	F1-score
Naïve bays	2	10	0.6087
Logistic regression	3	25	0.7605
Random forest	1	20	0.8178
Neural network	1	20	0.7788

Table 3. The confusion matrix of the Random forest where n_step = 1 and n_feature = 20

	Predicted : No	Predicted : Yes
Actual : No	1699	29
Actual : Yes	12	92

6 Discussion

In this paper, classification models to predict the likelihood of PM2.5 exceeding the healthy threshold are proposed. We implement 4 classification models including Naïve Bays, Logistic regression, Random forest and Neural network. These models achieve reasonably good F1-score ranging from 0.6 to 0.8 with their best set of parameters. Among different models and various parameters, the Random forest model with 1 time step (n_step = 1) utilizing top 20 features provides the best performance with 0.82 F1-score.

Compared to previous works, we include novel features in the models such as meteorological and air pollutants measured at different altitudes above the ground level. To evaluate a significance of features at the high level, F-values corresponding to these features are computed. Among top 30 features with high F-value, multiple features measured at high altitudes are selected. This implies an importance of including these high-level features into the models. According to the results, various features at different time steps play a role in the models due to their high F-value. Specifically, PM2.5 itself at previous time steps, NO2, NOX and SO2 pollutants are among the top important features.

The data set used as a case study is particularly collected from the air monitoring station located in Bangkok, Thailand. From the best of our knowledge, there is no previous work which conduct a study in the same direction as our work especially for the data set locally collected in Thailand. In this data set, a severity of an unbalance level between classes is relatively high. The re-sampling technique SMOTE is implemented to address this problem. Referring to our experiments, the models with SMOTE yield better performance than the ones without SMOTE. Even though only one case study is used to evaluate the performance of the classification models in this study, the methodology including machine learning models, features and relevant techniques can be applied and generalized to other data sets.

7 Conclusion

The proposed classification models aim to predict the likelihood of PM2.5 exceeding the pre-defined threshold. Meteorological data and air pollutants features are specifically considered at different levels above a ground along a vertical axis. Adding additional features which further draw insightful information from the data potentially enhances the predictive power of the models. Higher performance of the prediction can possibly be achieved if more complete and accurate data is available. With a larger data set, more complicated models such as deep learning approaches possibly enhance an accuracy of the prediction.

We are grateful to the Environmental Faculty, Kasetsart University for providing the meteorological data and air pollutants information.

References

1. L. Bai, J. Wang, X. Ma, H. Lu. Air pollution forecasts: An overview. Int. J. Environ. Res. Public Health, **15**, 780 (2018)
2. C. Bellinger, MS. Mohomed Jabbar, O. Zaiane, A. Osornio-Vargas. A systematic review of data mining and machine learning for air pollution epidemiology. BMC Public Health **17**, 907 (2017)
3. N.V. Chawla, K.W. Bowyer, L.O. Hall, W.P. Kegelmeyer. SMOTE: Synthetic Minority Over-sampling Technique. J. Artitif. Intell. Res. **16**, 321-357 (2002)
4. Y. Chu, Y. Liu, X. Li, Z. Liu, H. Lu, Y. Lu, Z. Mao, X. Chen, N. Li, M. Ren, F. Liu, L. Tian, Z. Zhu, H. Xiang. A Review on Predicting Ground PM2.5 Concentration Using Satellite Aerosol Optical Depth. Atmosphere **7**, 129 (2016)
5. N. Chuersuwan, S. Nimrat, S. Lekphet, T. Kerdkumrai. Levels and major sources of PM2.5 and PM10 in Bangkok Metropolitan Region. Environ. Int. **34**, 671-677 (2008)
6. W.G. Cobourn, An enhanced PM2.5 air quality forecast model based on nonlinear regression and back-trajectory concentrations. Atmos. Environ. **44**, 3015-3023 (2010)
7. L. Curtis, W. Rea, P. Smith-Willis, E. Fenyves, Y. Pan. Adverse health effects of outdoor air pollutants. Environ. Int. **32**, 815-830 (2006)
8. A.T. DeGaetano, O.M. Doherty. Temporal, spatial and meteorological variations in hourly PM2.5 concentration extremes in New York City. Atmos. Environ. **38**, 1547-1558 (2004)
9. M. Dong, D. Yang, Y. Kuang, D. He, S. Erdal, D. Kenski. PM2.5 concentration prediction using hidden semi-Markov model-based times series data mining. Expert Syst. Appl. **36**, 9046-9055 (2009)
10. H.K. Elminir. Dependence of urban air pollutants on meteorology. Sci. Total Environ. **350**, 225-237 (2005)
11. X. Feng, Q. Li, Y. Zhu, J. Hou, L. Jin, J. Wang. Artificial neural networks forecasting of PM2.5 pollution using air mass trajectory based geographic model and wavelet transformation. Atmos. Environ. **107**, 118-128 (2015)
12. Z. Haiming, S. Xiaoxiao. Study on Prediction of Atmospheric PM2.5 Based on RBF Neural Network. *2013 Fourth International Conference on Digital Manufacturing & Automation*. IEEE. 1287-1289 (2013)
13. T. Kanabkaew. Prediction of hourly particulate matter concentrations in Chiangmai, Thailand using MODIS Aerosol optical depth and ground-based meteorological data. Environ. Asia **6**, 65-70 (2013)
14. I.G. McKendry. Evaluation of Artificial Neural Networks for Fine Particulate Pollution (PM10 and PM2.5) Forecasting. J. Air Waste Assoc. **52**, 1096-1101 (2002)

15. D. Mishra, P. Goyal, A. Upadhyay. Artificial intelligence based approach to forecast PM2.5 during haze episodes: A case study of Delhi, India. Atmos. Environ. **102**, 239-248 (2015)
16. J.B. Ordieres, E.P. Vergara, R.S. Capuz, R.E. Salazar. Neural network prediction model for fine particulate matter (PM2.5) on the US–Mexico border in El Paso (Texas) and Ciudad Juárez (Chihuahua). Environ. Model. Softw. **20**, 547-559 (2005)
17. P. Perez, A. Trier, J. Reyes. Prediction of PM2.5 concentrations several hours in advance using neural networks in Santiago, Chile. Atmos. Environ. **34**, 1189-1196 (2000)
18. C.A. Pope, R.T. Burnett, M.J. Thun, E.E. Calle, D. Krewski, K. Ito, G.D. Thurston. Lung cancer, cardiopulmonary mortality, and long-term exposure to fine particulate air pollution. JAMA. **287**, 1132-1141 (2002)
19. W. Sun, H. Zhang, A. Palazoglu, A. Singh, W. Zhang, S. Liu. Prediction of 24-hour-average PM2.5 concentrations using a hidden Markov model with different emission distributions in Northern California. Sci. Total Environ. **443**, 93-103 (2013)
20. D. Voukantsis, K. Karatzas, J. Kukkonen, T. Rasanen, A. Karppinen, M. Kolehmainen. Intercomparison of air quality data using principal component analysis, and forecasting of PM10 and PM2.5 concentrations using artificial neural networks, in Thessaloniki and Helsinki. Sci. Total Environ. **409**, 1266-1276 (2011)
21. Q. Zhou, H. Jiang, J. Wang, J. Zhou. A hybrid model for PM2.5 forecasting based on ensemble empirical mode decomposition and a general regression neural network. Sci. Total Environ. 496, 264-274 (2014)
22. P. Amphanthong and P. Busababodhin. Forecasting PM10 in the upper northern area of Thailand with grey system theory. Burapha Science Journal. 20, 15-24 (2015)

Development of Environmentally Friendly Brake Lining Material

I K. Adi Atmika[1,*], *IDG. Ary* Subagia[1], *IW.* Surata[1], and *IN.* Sutantra[2]

[1]Mechanical Engineering Departement, Udayana University, Bali-Indonesia
[2]Mechanical Engineering Departement, Institut of Sepuluh Nopember, Surabaya-Indonesia

Abstract. Materials commonly used as brake pads are asbestos and alloys, but this material is very dangerous to the environment and health. This research was developed to answer these problems, namely to look for alternative brake pads that have good mechanical and structural properties and are environmentally friendly. Brake lining pads material is made from hybrid composite reinforced basalt, shells, alumina and bound using phenolic resin polymer (PR-51510i). This brake pads material is produced through a sintering process with an emphasis of 2000 kg for 30 minutes at a fixed temperature of 160°C. This hybrid composite is made in as many as five variations, each of which is tested for wear resistance using a pin on disc test based on ASTM G 99-95a standards, while destilled water absorption test was based on ASTM D 570-98. The greatest wear rate is 0.000090 g/cm, which is still lower than wear rate of asbestos brake pad materials, and the highest destilled water absorption of the brake pads specimens obtained was 0.041558 still lower than the destilled water absorption of asbestos brake pads.

1 Indroduction

Brake lining pads is part of the braking system that immediately rubs against the rotating part. In general, brake pad material is made of asbestos material because its performance is still good to a temperature of 800°C [1], [2]. However, asbestos ingredients have been discontinued because they have carcinogenic properties that have a negative impact on the environment and human health [3], [4].

Then many researchers developed asbestos-free brake pad material. In 2014, Mgbemena et al developed the Palm kernel shell (PKS) based Pulverized non-asbestos friction material for reinforcing constituents, controlling friction imparting, elastomeric additives, fire retarding components and thermosetting resin. The results showed that the degradation temperature of the palm was obtained at 53.84 C with a final degradation temperature of 634.87 C and a percentage weight loss of 86%.[5]. Then other researchers also developed shells based brake pads with a grit size of 600 μm. The material is tested still at speeds below 100 km / hr [6]. Yawas et al. performed morphological tests of shellfish granules for friction plate material. The development of asbestos-free automotive brakes using periwinkle shell particles as frictional filler material is presented. This was with the characteristics of the periwinkle shell, which is mostly deposited as waste, in replacing asbestos which has been found to be carcinogenic [7], and the thermal test of the shellfish material for scraping material has also been carried out and has quite good properties [5]. But the results achieved at this time have not been able to maintain mechanical properties, especially against wear resistance and material performance at high temperatures.

This paper presents the characteristics of wear resistance and friction coefficient of brake pads material made from hybrid composite reinforced basalt, shells, alumina and bound using phenolic resin matrix polymer (PR-51510i). Basalt stone is one of the volcanic eruption materials that has heat resistance up to 1500°C [8], corrosion resistant, low water absorption and resistant to chemical treatment and non-toxic [9]. Basalt stone properties have excellent physical and mechanical properties, high tenacity, and high wear resistance [10], and can replace glass fiber [11]. Then the most important properties of this material are to have low thermal conductivity.

2 Material and Method

This research was carried out by hybridizing 3 types of material as reinforcement and one material as a matrix. The reinforcing materials are basalt powder, shellfish powder and alumina powder in the form of solid particles with a size of 60 mesh, then as a matrix material composite phenolic resin (PR-51510i) is used. Basalt properties are shown in table 1. The shellfish powder material consists of 66.70% CaO, 22.28% MgO, 7.88% SiO_2, 1.25% Al_2O_3 and 0.03% Fe_2O_3. The process of making specimens is carried out by mixing reinforcing material and matrix material in the dry phase. Variations in the weight fraction of the brake pad material studied are shown in table 2.

The specimen formation process uses the sintering casting method at 160^0C, a pressure of 3 tons for 30 minutes. Testing of 5 specimen variations and 1 specimen from asbestos material was carried out by pin on disc

* Corresponding author: tutadi2001@yahoo.com

method based on ASTM G 99-95a standard (Figure 1 and Figure 2). Each specimen sample was given a load of 36 N taken in 83.3 minutes or 200 meters with an engine speed of 120 rpm which resulted in friction on the surface of the sample. While the destilled water absorption test was based on ASTM D 570-98, then asbestos brake pads (X) is tested with the same method and standard.

Table 1. Mechanical properties of basalt material

Properties of Basalt	Value (unity)
Density	2600-2630 (kg/m^3)
Tensile strength	500k-550k (psi)
Sintering Temperatur	1050 (°C)
Operation Temperatur	-265-+700 (°C)
Modulus of elastisitas	9100-1100 (kg/mm^3)
Melting point	1170(°C)
Heat resistance	700-1000 (deg.C)
Elongation at break	3.15(%)

Table 2. Weight fraction ratio of hybrid composite

Composition	Basalt powder (%)	Shellfish powder (%)	Alumina powder (%)	Phenolic Resin (%)
A	45	5	10	40
B	40	10	10	40
C	35	15	10	40
D	30	20	10	40
E	25	25	10	40

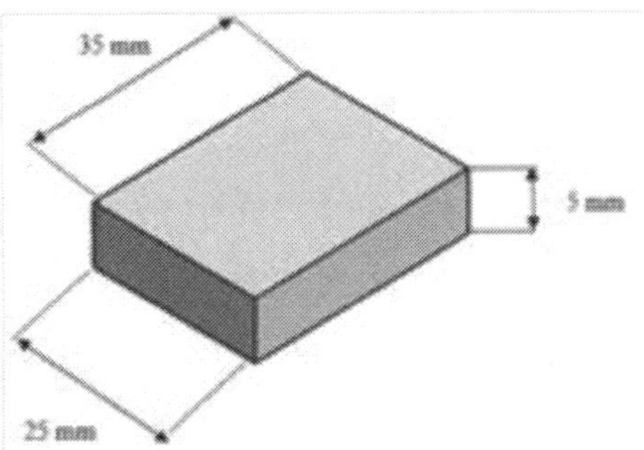

Fig. 1. Specimen geometry wear test

Fig.2. Pin on disc Equipment

3 Result and Discussion

The wear rate and friction coefficient (μs) results are shown in table 3. Then the comparison of wear rate for the five hybrid composite variations being developed and the asbestos (X) brake lining pad materials is shown in Figure 3 and Figure 4. While the results of destilled water absorption tests are shown in table 4. Then the destilled water absorption distribution of the five hybrid composite variations being developed compared to asbestos brake lining pads material (X) are shown in Figure 5.

Table 3. Friction coefficients and wear rate values

Composition	μs	wear rate (g/cm)
A	0,410	0,000071
B	0,402	0,000070
C	0,398	0,000080
D	0,370	0,000090
E	0,360	0,000090
X	0,330	0,000110

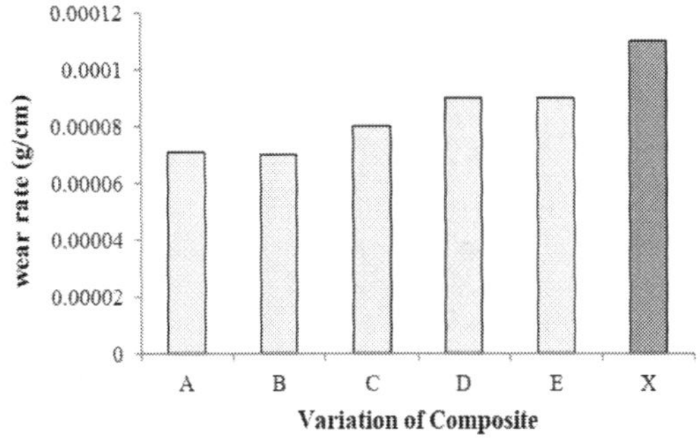

Fig.3. Wear rate of the hybrid composite and asbestos

Figure 3 shows a relationship diagram between variations in specimens with wear rate values, the lowest 0.000071 g/cm and the highest is 0.00009 g/cm. The reduced amount of basalt content in specimens of wear rate value tends to be even greater, but still smaller than the wear rate value of asbestos brake pads specimens. So the five specimen variations of hybrid compsite brake pads have better wear rate value compared to asbestos brake pads.

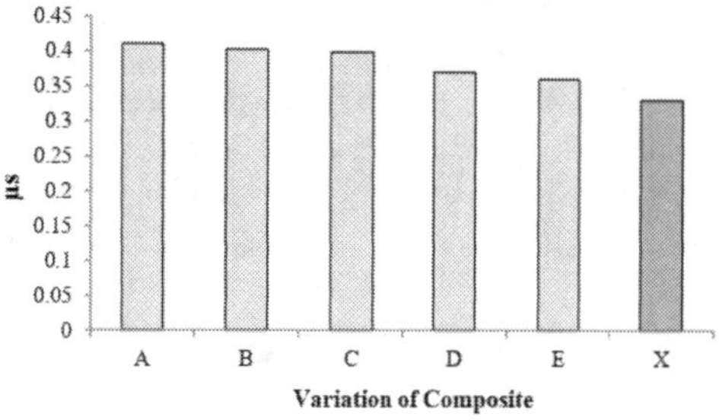

Fig.4. coefficient of friction from hybrid composite and asbestos

Figure 4 shows the average friction coefficient of hybrid composite brake pads for each composition variation. The friction coefficient value obtained thoroughly is higher than the friction coefficient value of asbestos brake pads. The difference in friction coefficient from hybrid composite brake pads is 9% to 25%.

Table 4: destilled water absorption

Variation of Composite	destilled water absorption (%)				
	3 days	14 days	24 days	30 days	60 days
X	0.032136	0.041588	0.043478	0.043478	0.043478
A	0.038961	0.041558	0.041558	0.041558	0.041558
B	0.031414	0.036649	0.036649	0.036649	0.036649
C	0.028721	0.033943	0.036554	0.036554	0.036554
D	0.026178	0.028796	0.031414	0.031414	0.031414
E	0.023499	0.026110	0.028721	0.028721	0.028721

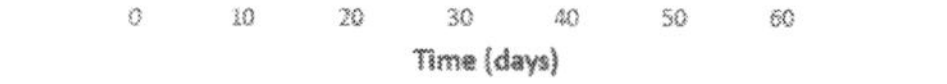

Fig.5. Destilled water absorption of the specimens

Figure 5 shows an overview of trends in destilled water absorption by hybrid composite materials being developed as well as asbestos brake pads material. It is seen that the trend after 24 days of immersion began to be constant for each specimen variation. The lowest destilled water absorption of specimens E was 0.028721% compared to the other four specimens and all of them were still below the destilled water absorption capacity of the asbestos brake pads material (X) [12].

Based on these results, basalt as the main reinforcing material from hybrid composite brake pads has had a quite effective effect on wear resistance, where with increasing basalt content the brake pads hardness also increases [13]. Furthermore, shellfish powder also affects the increase in friction coefficient that occurs in hybrid composite brake pads. While the alumina particles with a fixed amount due to the sintering temperature that has been carried out have not yet been achieved, it has not given a significant effect on the characteristics and friction coefficient value of composite hybrid brake pads (Figure 6).

Fig.6. SEM photos of composites developed (2000x)

From Figure 6, observations of photography of alumina particles have not been bound by the composite matrix phenolic resin properly [14].

4 Conclucions

Reducing the percentage of basalt in each composition results in increased wear rate values, whereas the value of the coefficient of friction in each composition decreases. The wear rate value of the hybrid composite material is still smaller than the asbestos brake pad material, as well as the higher friction coeficient compared to the friction coefficient of the asbestos brake pad material. Furthermore the reduction in the percentage of basalt in each composition results in a decrease in destilled water absorption and is still lower than the absorption capacity of asbestos brake lining pads. So the hybrid composite material developed is very possible to replace asbestos material as an environmentally friendly brake pad materials.

References

1. M. A. Sivarao, M. S. Rizal, and A. Kamely, "An investigation toward development of economical brake lining wear alert system," *Int. J. Eng. Technol. IJET*, vol. 9, no. 9, pp. 251–256, 2009.
2. N. M. Kinkaid, O. M. O'Reilly, and P. Papadopoulos, "Automotive disc brake squeal," *J. Sound Vib.*, vol. 267, no. 1, pp. 105–166, 2003.
3. M. G. Jacko, P. H. S. Tsang, and S. K. Rhee, "Automotive friction materials evolution during the past decade," *Wear*, vol. 100, no. 1–3, pp. 503–515, 1984.

4. L. St, "US Survey shows imports of asbestos brake materials increasing," *St. Louis*, 2004.
5. C. O. Mgbemena, C. E. Mgbemena, and M. O. Okwu, "Thermal stability of pulverized palm kernel shell (PKS) based friction lining material locally developed from spent waste," *ChemXpress*, vol. 5, pp. 115–122, 2014.
6. P. C. Olele, A. C. Nkwocha, I. C. Ekeke, M. O. Ileagu, and E. O. Okeke, "Assessment of Palm Kernel Shell as Friction Material for Brake Pad Production," *Int. J. Eng. Manag. Res.*, vol. 6, no. 1, pp. 281–284, 2016.
7. D. S. Yawas, S. Y. Aku, and S. G. Amaren, "Morphology and properties of periwinkle shell asbestos-free brake pad," *J. King Saud Univ. Sci.*, vol. 28, no. 1, pp. 103–109, 2016.
8. V. Dhand, G. Mittal, K. Y. Rhee, S. J. Park, and D. Hui, "A short review on basalt fiber reinforced polymer composites," *Compos. Part B Eng.*, vol. 73, no. December, pp. 166–180, 2015.
9. V. Fiore, G. Di Bella, and A. Valenza, "Glass–basalt/epoxy hybrid composites for marine applications," *Mater. Des.*, vol. 32, no. 4, pp. 2091–2099, 2011.
10. A. A. Dalinkevich, K. Z. Gumargalieva, S. S. Marakhovsky, and A. V Soukhanov, "Modern basalt fibrous materials and basalt fiber-based polymeric composites," *J. Nat. Fibers*, vol. 6, no. 3, pp. 248–271, 2009.
11. V. Lopresto, C. Leone, and I. De Iorio, "Mechanical characterisation of basalt fibre reinforced plastic," *Compos. Part B Eng.*, vol. 42, no. 4, pp. 717–723, 2011.
12. N. Venkateshwaran, A. ElayaPerumal, A. Alavudeen, and M. Thiruchitrambalam, "Mechanical and water absorption behaviour of banana/sisal reinforced hybrid composites," *Mater. Des.*, vol. 32, no. 7, pp. 4017–4021, 2011.
13. R. D. Hyndman and M. J. Drury, "Physical properties of basalts, gabbros, and ultramafic rocks from DSDP Leg 37," *DSDP*, vol. 37, pp. 395–401, 1977.
14. S. Choi and J. Kim, "Thermal conductivity of epoxy composites with a binary-particle system of aluminum oxide and aluminum nitride fillers," *Compos. Part B Eng.*, vol. 51, pp. 140–147, 2013.

AUTHOR INDEX